高等学校"十三五"规划教材

机电工程专业英语

English in Mechatronic Engineering

（第 10 版）

主编　施平

哈尔滨工业大学出版社

内 容 提 要

本书以培养学生专业英语能力为主要目标。全书共分六个部分,主要内容为机械零件和设计,机床和加工,质量和生产率,制造工程和自动化,现代制造技术及其发展、教育。本书具有较强的实用性和知识延伸性。本书既可作为高等学校机电工程、机械设计制造及自动化、机械工程及自动化等专业学生的教材,也可供从事上述专业的工程技术人员学习、参考之用。

图书在版编目(CIP)数据

机电工程专业英语/施平主编. —10 版. —哈尔滨:哈尔滨工业大学出版社,2017.7(2022.6 重印)
ISBN 978-7-5603-6714-9

Ⅰ.①机… Ⅱ.①施… Ⅲ.①机电工程-英语-高等学校-教材 Ⅳ.①TH

中国版本图书馆 CIP 数据核字(2017)第 147351 号

责任编辑	张秀华	
封面设计	卞秉利	
出版发行	哈尔滨工业大学出版社	
社　　址	哈尔滨市南岗区复华四道街 10 号　邮编 150006	
传　　真	0451-86414749	
网　　址	http://hitpress.hit.edu.cn	
印　　刷	哈尔滨久利印刷有限公司	
开　　本	880mm×1230mm　1/32　印张 11.25　字数 360 千字	
版　　次	2017 年 7 月第 10 版　2022 年 6 月第 3 次印刷	
书　　号	ISBN 978-7-5603-6714-9	
定　　价	30.00 元	

(如因印装质量问题影响阅读,我社负责调换)

前　　言

英语作为一门主要的国际交流语言,其作用已日显重要。为了更快、更准确地了解本专业的国际发展动向,学习和借鉴国外的先进技术和管理经验,专业英语的应用能力已成为高等院校学生和科技工作者应该具备的素质之一。

编写本书的主要目的是帮助机电工程、机械设计制造及自动化、机械工程及自动化等专业的学生们提高专业英语的实际应用能力。本书初版于 1996 年 5 月,这次是在 2010 年第 8 版的基础上又做了全面修订。

全书共分六个部分。第一部分为机械零件和设计,第二部分为机床和加工,第三部分为质量和生产率,第四部分为制造工程和自动化,第五部分为现代制造技术及其发展,第六部分为教育。

课文内容比较新颖,文体规范,难度适中。为了适应专业英语教学的要求,书中内容既对学生学过的课程进行了必要的覆盖,又有所拓宽和延伸,力求反映机电工程和机械制造技术的现状和发展趋势,既可提高读者英语阅读水平,又能使读者了解学科前沿。

本书由施平主编,参加编写工作的有梅雪、田锐、胡明、李越、施晓东、魏思欣、侯双明,由贾艳敏担任主审。对书中的不足之处,恳请广大读者批评指正。

编　者
2014 年 10 月

CONTENTS

1 MACHINE COMPONENTS AND DESIGN

1. Couplings, Clutches, and Springs ·· (1)
2. Belts, Chains, and Brakes ··· (5)
3. Rolling Bearings ·· (11)
4. Shafts and Splines ··· (16)
5. Threaded Fasteners ··· (20)
6. Lubrication ·· (25)
7. Machine Tool Frames ··· (30)
8. Engineering Graphics ··· (35)
9. Sectional Views ··· (40)
10. Engineering Design ·· (47)
11. Engineering Design and Safety Factors ····························· (52)
12. Computer Applications in Design and Graphics ·················· (58)
13. Responsibility, Liability, and Litigation ···························· (63)

2 MACHINE TOOLS AND MACHINING

14. Engine Lathes ··· (68)
15. Milling Operations ··· (73)
16. Drilling Operations ·· (79)
17. Grinding Machines ··· (85)
18. Machining ·· (91)
19. Numerical Control ··· (94)

20. Machine Tool Motors ……………………………… (100)
21. Numerical Control ……………………………… (105)
22. Choice of Manual or CNC Machine Tools …………… (110)
23. Hard-Part Machining with Ceramic Inserts …………… (114)
24. Nontraditional Manufacturing Process ………………… (119)
25. Design Considerations for NC Machine Tools ………… (124)
26. Laser-Assisted Machining and Cryogenic Machining Technique
    ……………………………………………………… (129)

3 QUALITY AND PRODUCTIVITY

27. Quality in Manufacturing ……………………………… (133)
28. Quality in the Modern Business Environment ………… (137)
29. Design and Manufacturing Tolerances ………………… (142)
30. Product Reliability Requirements …………………… (146)
31. Product Reliability ……………………………… (150)
32. Total Quality Management ……………………………… (154)
33. Employees Take Charge of Their Jobs ………………… (158)
34. Accelerating Development with Acquired Technology ………
    ……………………………………………………… (165)

4 MANUFACTURING AND AUTOMATION

35. Careers in Manufacturing ……………………………… (169)
36. Numerical Control Software ……………………………… (173)
37. Computers in Manufacturing ……………………………… (177)
38. Automated Assembly Equipment Selection …………… (181)
39. Motion Control Advances Assembly …………………… (186)
40. Versatility of Modular Fixturing ……………………… (191)
41. Product Design for Manufacture and Assembly ……… (195)
42. Making a Cost Estimate ……………………………… (201)
43. Determining the Cost of a Product …………………… (204)

5 MODERN MANUFACTURING ENGINEERING AND DEVELOPMENT

- 44. Industrial Robots .... (210)
- 45. Robotic Sensors .... (215)
- 46. Types of Robots .... (220)
- 47. Coordinate Measuring Machines .... (223)
- 48. Computer Aided Process Planning .... (227)
- 49. Computer Numerical Control .... (231)
- 50. AI: Promises Start to Pay Off .... (234)
- 51. Online Design .... (238)
- 52. Flexible Manufacturing Systems .... (243)
- 53. Computer-Integrated Manufacturing .... (246)
- 54. Mechanical Engineering in the Information Age .... (250)
- 55. Mechatronics .... (255)

6 EDUCATION

- 56. Manufacturing Engineering Education .... (260)
- 57. Manufacturing Research Centers at U.S. Universities ... (267)
- 58. A New Engineering Course .... (271)
- 59. Online Web-Based Learning .... (277)
- 60. Laboratory Handbook for Technical Reporting [Ⅰ] Introduction .... (282)
- 61. Laboratory Handbook for Technical Reporting [Ⅱ] Effective Report Writing .... (285)
- 62. Laboratory Handbook for Technical Reporting [Ⅲ] Technical Report Elements .... (290)

Glossary .... (296)
主要参考文献 .... (351)

1 MACHINE COMPONENTS AND DESIGN

1. Couplings, Clutches, and Springs

A coupling is a device for connecting the ends of adjacent shafts. In machine construction, couplings are used to effect a semipermanent connection between adjacent rotating shafts. The connection is permanent in the sense that it is not meant to be broken during the useful life of the machine, but it can be broken and restored in an emergency or when worn parts are replaced.

There are several types of shaft couplings, their characteristics depend on the purpose for which they are used. If an exceptionally long shaft is required for a line shaft in a manufacturing plant or a propeller shaft on a ship, it is made in sections that are coupled together with rigid couplings(see Fig. 1.1).

Figure 1.1 A rigid coupling

In connecting shafts belonging to separate devices (such as an electric motor and a gearbox), precise aligning of the shafts is difficult and a flexible coupling is used (see Fig. 1.2). This coupling connects the shafts in such a way as to minimize the harmful effects of shaft misalignment. Flexible couplings also permit the shafts to deflect under their separate systems of loads and to move freely (float) in the axial direction without interfering with one another. Flexible couplings can also serve to reduce the intensity of shock loads and vibrations transmitted from one shaft to another.

Figure 1.2　A flexible coupling

A clutch (see Fig. 1.3) is a device for quickly and easily connecting or disconnecting a rotatable shaft and a rotating coaxial shaft. Clutches are usually placed between the input shaft to a machine and the output shaft from the driving motor, and provide a convenient means for starting and stopping the machine and permitting the driver motor or engine to be started in an unloaded state.

Figure 1.3　A clutch

The rotor (rotating member) in an electric motor has rotational inertia, and a torque is required to bring it up to speed when the motor is started. If the motor shaft is rigidly connected to a load with a large rotational inertia, and the motor is started suddenly by closing a switch, the motor may not have sufficient torque capacity to bring the motor shaft up to speed before the windings in the motor are burned out by the excessive current demands. A clutch between the motor and the load shafts will restrict the starting torque on the motor to that required to accelerate the rotor and parts of the clutch only.

On some machine tools it is convenient to let the driving motor run continuously and to start and stop the machine by operating a clutch. Other machine tools receive their power from belts driven by pulleys on intermediate shafts that are themselves driven by belts from long lineshafts that serve a group of machines.

A spring is a load-sensitive, energy-storing device, the chief characteristics of which are an ability to tolerate large deflections without failure and to recover its initial size and shape when loads are removed. The three major classifications of springs are compression, extension, and torsion (see Fig. 1.4). Although most springs are mechanical and derive their effectiveness from the flexibility inherent in metallic elements, hydraulic springs and air springs are also obtainable.

(a) Helical compression spring (b) Helical extension spring (c) Helical torsion spring

Figure 1.4　Three types of springs

Springs are used for a variety of purposes, such as supplying the motive power in clocks and watches, cushioning transport vehicles, measuring weights, restraining machine elements, mitigating the transmission of periodic disturbing forces from unbalanced rotating machines to the supporting structure, and providing shock protection for delicate instruments during shipment.

Words and Expressions

coupling ['kʌpliŋ] n. 耦合,联轴器,连接器
clutch [klʌtʃ] n. 离合器; v. 使离合器接合
shaft [ʃɑːft] n. 轴,辊
semipermanent [ˌsemiˈpəːmənənt] a. 半永久性的,暂时的
in the sense that... 在…意义上
shaft coupling 联轴器
exceptionally [ikˈsepʃənli] ad. 格外地,特别地
line shaft 动力轴,主传动轴
propeller [prəˈpelə] n. 螺旋桨,推进器
couple ['kʌpl] v. 使联在一起,联接,力偶
rigid coupling 刚性联轴器,刚性联接
aligning [əˈlainiŋ] n. 找正,直线对准
flexible coupling 弹性(挠性)联轴器
misalignment [ˈmisəlainmənt] n. 未对准,轴线不重合,安装误差
deflect [diˈflekt] v. 偏移,弯曲,下垂
axial [ˈæksiəl] a. 轴的,轴向的
interfere [ˌintəˈfiə] v. 干涉,干扰,同…抵触,冲突(with)
shock [ʃɔk] n. 冲击,碰撞
shock load 冲击载荷,突加载荷
vibration [vaiˈbreiʃən] n. 振动
coaxial [kəuˈæksiəl] a. 同轴的,共轴的
means [ˈmiːnz] n. 手段,方法; v. 意味,想要
driver motor 主驱动电动机
rotor [ˈrəutəː] n. 转子,电枢,转动体

rotational inertia 转动惯量
torque [tɔ:k] n. 转矩,扭矩; v. 扭转
winding ['waindiŋ] n. 绕组,线圈
burn out 烧坏,烧掉
machine tool 机床
current 电流
pulley ['puli] n. 滑轮,皮带轮
intermediate shaft 中间轴
deflection [di'flekʃən] n. 偏移,偏转,弯曲,挠度
derive [di'raiv] v. 从…得到,获得,引伸出
inherent in 为…所固有,固有的
metallic [mi'tælic] a. 金属的
hydraulic [hai'drɔ:lik] a. 液压的
motive power 动力,原动力
cushion ['kuʃən] n. 缓冲器; v. 缓冲,减振
restrain [ris'trein] v. 抑制,约束,限制
mitigate ['mitigeit] v. 使缓和,减轻,防止
disturbing force 干扰力
delicate ['delikit] a. 精密的,精巧的,灵敏的

2. Belts, Chains, and Brakes

Belts are used to transmit power from one shaft to another smoothly, quietly, and inexpensively. Belts are frequently necessary to reduce the higher rotative speeds of electric motors to the lower values required by mechanical equipment. Chains provide a convenient and effective means for transferring power between parallel shafts. The function of the brake is to turn mechanical energy into heat.

There are four main belt types: flat, round, V, and synchronous. A widely used type of belt, particularly in industrial drives and vehicular applications, is the V-belt drive. The V shape causes the belt to wedge tightly into the groove of the sheave, increasing friction and allowing high torques to be transmitted before slipping occurs. Since the cost of V-belts is relatively low, the power output of a V-belt system may be increased by operating several belts side by side (see Fig. 2.1). Another type of V-belt drive is the poly V-belt drive (see Fig. 2.2). V-belt dive helps protect the machinery from overload, and it damps and isolates vibration.

Figure 2.1   V-belts and multiple V-belt drive

Figure 2.2   Poly V-belt drive

Other belts, such as flat belts, are used for long center distances and high speed applications. A flat belt drive has an efficiency of around 98%, which is nearly the same as for a gear drive. A V-belt drive can transmit more power than a flat belt drive. However, the efficiency of a V-belt drive varies between 70 and 96%.

Synchronous belts (also known as timing belts) have evenly spaced teeth on the inside circumference (see Fig. 2.3). Synchronous belts do not slip and hence transmits power at a constant angular velocity ratio.

Figure 2.3 Synchronous belt drive

However, if very large ratios of speed reduction are required in the drive, gear reducers are desirable because they can typically accomplish large reductions in a rather small package. The output shaft of the gear-type speed reducer is generally at low speed and high torque. If both speed and torque are satisfactory for the application, it could be directly coupled to the driven machine.

Chain drives combine some of the more advantageous features of belt and gear drives. Chain drives provide almost any speed ratio. Their chief advantage over gear drives is that chain drives can be used with arbitrary center distances. Compared with belt drives, chain drives offer the advantage of positive (no slip) drive and therefore greater power capacity. An additional advantage is that not only two but also many shafts can be driven by a single chain at different speeds, yet all have synchronized motions. Primary applications are in conveyor systems, farm machinery, textile machinery, and motorcycles.

In chain drive applications, toothed wheels called sprockets mate with a chain to transmit power from one shaft to another (see Fig. 2.4). The most common type of chain is the roller chain (see Fig. 2.5), in which the roller on each bushing provides exceptionally low friction between the chain and the sprockets. Of its diverse applications, the most familiar one is the roller chain drive on a bicycle. A roller chain is generally made of hardened steel. Sprockets are generally made of steel or cast iron, but some applications use aluminum alloy or plastic. Nevertheless, stainless steel and bronze chains are obtainable where corrosion resistance is needed.

Figure 2.4 Chain drives

Figure 2.5 Portion of a roller chain

A brake is similar to a clutch except that one of the shafts is replaced by a fixed member. A brake is a mechanical device which inhibits motion (see Fig. 2.6). The basic function of a brake is to absorb energy (i. e., to convert kinetic and potential energy into friction heat) and to dissipate the resulting heat without developing destructively high temperatures. Clutches also absorb energy and dissipate heat, but usually at a lower rate. Where brakes are used more or less continuously for extended periods of time, provision must be made for rapid transfer of heat to the surrounding atmosphere. For intermittent operation, the thermal capacity of the parts may permit much of the heat to be stored, and then dissipated over a longer period of time. Brake and clutch parts must be designed to avoid objectionable thermal stresses and thermal distortion.

The rate at which heat is generated on a unit area of friction

MACHINE COMPONENTS AND DESIGN 9

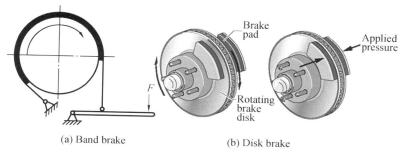

(a) Band brake (b) Disk brake

Figure 2.6 Brakes

interface is equal to the product of the normal pressure, coefficient of friction, and rubbing velocity. Manufacturers of brakes and of brake lining materials have conducted tests and accumulated experience enabling them to arrive at empirical values of pV (normal pressure times rubbing velocity) and of power per unit area of friction surface that are appropriate for specific types of brake design, brake lining material, and service conditions.

Words and Expressions

belt drive 带传动(由柔性带和带轮组成传递运动和动力的机械传动,分摩擦传动和啮合传动)

chain drive 链传动(利用链与链轮轮齿的啮合来传递动力和运动的机械传动)

rotative ['rəutətiv] *a*. 回转的,转动的

brake [breik] *n*. 制动器(具有使运动部件减速、停止或保持停止状态等功能的装置)

vehicular [vi'hikjulə] *a*. 车辆的,运载工具的

flat belt 平带(横截面为矩形或近似为矩形的传动带,其工作面为宽平面)

round belt 圆带(横截面为圆形或近似为圆形的传动带)

V-belt V带(横截面为等腰梯形或近似为等腰梯形的传动带,其工作面为两侧面)

synchronous ['siŋkrənəs] a. 同时的,同步的

synchronous belt 同步带(横截面为矩形或近似为矩形,内表面或者内、外表面具有等距横向齿的环形传动带)

wedge [wedʒ] n. 楔; v. 楔入,楔进

sheave [ʃi:v] n. V带轮,有槽的带轮,即 grooved pulley

multiple V-belt 多条V带,多根V带

poly V-belt 多楔带(以平带为基体,内表面具有等距纵向楔的环形传动带)

damp [dæmp] v. 阻尼,使衰减,抑止

timing belt 同步齿形带

conveyor [kən'veiə] n. 输送机(具有输送功能的机械)

roller chain 滚子链(组成零件中具有回转滚子,且滚子表面在啮合时直接与链轮齿接触的链条)

positive (no slip) drive 强制(无滑动)传动

sprocket ['sprɔkit] n. 链轮

bushing ['buʃiŋ] n. 套筒

hardened steel 淬硬钢,淬火钢

dissipate ['disipeit] v. 消散

thermal distorsion 热变形

band brake 带式制动器(用制动带的内侧面作为摩擦副接触面的制动器)

disk brake 盘式制动器(用圆盘的端面作为摩擦副接触面的制动器)

brake lining 制动衬片,摩擦片,制动衬带

brake pad 摩擦衬块,制动衬块

brake disk 制动盘(以端平面为摩擦工作面的圆盘形运动部件)

product 乘积

normal pressure 正压力

rubbing velocity 摩擦速度

service condition 使用条件,工作条件,使用状态

3. Rolling Bearings

Rolling bearings can carry radial, thrust or combination of the two loads. Accordingly, most rolling bearings are categorized in one of the three groups: radial bearings for carrying loads that are primarily radial, thrust bearings for supporting loads that are primarily axial, and angular contact bearings or tapered roller bearings for carrying combined radial and axial loads. Figure 3.1a shows a common single-row, deep groove ball bearing. The bearing consists of an inner ring, an outer ring, the balls and the separator. To increase the contact area and hence permit larger loads to be carried, the balls run in curvilinear grooves in the rings called raceway. The radius of the raceway is very little larger than the radius of the ball. This type of bearing can stand a radial load as well as some thrust load. Some other types of rolling bearings are shown in Figs. 3.1b and 3.1c.

(a) Deep groove ball bearing   (b) Thrust ball bearing   (c) Needle roller bearing

Figure 3.1 Some types of rolling bearings

In high speed machining centers, rolling bearings are used as spindle bearings. High speed spindle bearings available today include roller, tapered roller, and angular contact ball bearings (Fig. 3.2). Selection criteria depend on the spindle specifications and the speed needed for metalcutting.

(a) Roller bearing　　(b) Tapered roller bearing　　(c) Angular contact ball bearing

Figure 3.2　Some spindle bearings

Angular contact bearings are the type most commonly used in very high speed spindle design. These bearings provide precision, load carrying capacity, and the speed needed for metalcutting. Angular contact ball bearings have a number of precision balls fitted into a precision steel race, and provide both axial and radial load carrying capacity when properly preloaded.

In some cases, tapered roller bearings are used because they offer higher load carrying capacity and greater stiffness than ball bearings. However, tapered roller bearings do not allow the high speeds required by many spindles.

Angular contact ball bearings are available with a choice of preloading magnitude, typically designated as light, medium, and heavy. Light preloaded bearings allow maximum speed and less stiffness. These bearings are often used for very high speed applications, where cutting loads are light.

Heavy preloading allows less speed, but higher stiffness. To provide the required load carrying capacity for a metalcutting machine tool spindle, several angular contact ball bearings are used together. In this way, the bearings share the loads, and increase overall spindle stiffness.

Hybrid ceramic bearings (see Fig. 3.3) are a recent

development in bearing technology that uses ceramic (silicon nitride) material to make precision balls (see Fig. 3.4). The ceramic balls, when used in an angular contact ball bearing, offer distinct advantages over typical bearing steel balls.

Figure 3.3　Hybrid ceramic bearing

Figure 3.4　Silicon nitride balls

Ceramic balls have 60% less mass than steel balls. This is significant because as a ball bearing is operating, particularly at high rotational speeds, centrifugal forces push the balls to the outer race, and even begin to deform the shape of the ball. This deformation leads to rapid wear and bearing deterioration. Ceramic balls, with less mass, will not be affected as much at the same speed. In fact, the use of ceramic balls allows up to 30% higher speed for a given ball bearing size, without sacrificing bearing life.

Due to the nearly perfect roundness of the ceramic balls, hybrid ceramic bearings operate at much lower temperatures than steel ball bearings, which results in longer life for the bearing lubricant. Tests show that spindles utilizing hybrid ceramic bearings exhibit higher rigidity and have higher natural frequencies, making them less sensitive to vibration.

Bearing lubrication is necessary for angular contact ball bearings to operate properly. The lubricant provides a microscopic film between the rolling elements to prevent abrasion. In addition, it protects surfaces from corrosion, and protects the contact area from particle contamination.

Grease is the most common and most easily applied type of lubricant. It is injected into the space between the balls and the races, and is permanent. Grease requires minimal maintenance. Generally, high speed spindles utilizing grease lubrication do not allow replacement of the grease between bearing replacements. During a bearing replacement, clean grease is carefully injected into the bearing.

Often at high rotational speeds, lubrication with grease is not sufficient. Oil is then used as a lubricant, and delivered in a variety of ways. Maintenance of the lubrication system is vital, and must be closely monitored to ensure that proper bearing conditions are maintained. Also, use of the correct type, quantity, and cleanliness of lubricating oil is critical.

Words and Expressions

radial bearing 向心轴承（主要用于承受径向载荷的滚动轴承）

thrust bearing 推力轴承（主要用于承受轴向载荷的滚动轴承）

angular contact ball bearing 角接触球轴承

tapered roller bearing 圆锥滚子轴承

deep groove ball bearing 深沟球轴承

inner ring 内圈（滚道在外表面的轴承套圈）

outer ring 外圈（滚道在内表面的轴承套圈）

separator ['sepəreitə] n. （轴承）保持架

curvilinear [kə:vi'liniə] a. 曲线的，由曲线而成的

raceway ['reiswei] n. （轴承的）滚道

thrust ball bearing 推力球轴承（滚动体是球的推力滚动轴承）

needle roller bearing 滚针轴承（滚动体是滚针的向心滚动轴承）

machining center 加工中心

spindle ['spindl] n. 轴，（机床）主轴

spindle bearing 主轴轴承

rotational speed 转速

race （轴承的）滚道

preload ['pri:'ləud] n.; vt. 预载荷，在施加"使用"载荷（外部载荷）前，通过相对于另一轴承的轴向调整而作用在轴承上的力，或由轴承内滚道与滚动体的尺寸形成"负游隙"（内部预载荷）而产生的力。

cutting tool 刀具

specification [ˌspesifi'keiʃən] n. 规范，规格，说明书，[pl.] 技术参数，技术要求

load carrying capacity 承载能力

metalcutting 金属切削

hybrid ceramic bearing 混合陶瓷轴承，复合陶瓷轴承

silicon nitride 氮化硅

roundness ['raundnis] *n.* 圆度
cutting load 切削载荷
natural frequency 固有频率
abrasion [ə'breiʒən] *n.* 磨损,磨耗,磨损处
grease [griːs] *n.* 脂,润滑脂
lubrication [ˌluːbri'keiʃən] *n.* 润滑,润滑作用
cleanliness ['klenlinis] *n.* 清洁度

4. Shafts and Splines

Virtually all machines contain shafts. The most common shape for shafts is circular and the cross section can be either solid or hollow (hollow shafts can result in weight savings). Rectangular shafts are sometimes used, as in slotted screw driver blades (see Fig. 4.1).

Figure 4.1　A slotted screw driver

A shaft must have adequate torsional strength to transmit torque (see Fig. 4.2) and not be over stressed. It must also be torsionally stiff enough so that one mounted component does not deviate excessively from its original angular position relative to a second component mounted on the same shaft. Generally speaking, the angle of twist should not exceed one degree in a shaft length equal to 20 diameters.

Shafts are mounted in bearings and transmit power through such devices as gears, pulleys, cams and clutches. These devices introduce forces which attempt to bend the shaft (see Fig. 4.3); hence, the shaft must be rigid enough to prevent overloading of the supporting bearings. In general, the bending deflection of a shaft should not exceed 0.5 mm per meter of length between bearing supports.

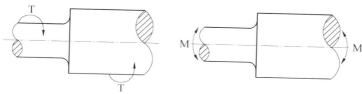

Fig. 4.2 Torsion of stepped shaft   Fig. 4.3 Bending of stepped shaft

In addition, the shaft must be able to sustain a combination of bending and torsional loads. Thus an equivalent load must be considered which takes into account both torsion and bending. Also, the allowable stress must contain a factor of safety which includes fatigue, since torsional and bending stress reversals occur.

For diameters less than 75 mm, the usual shaft material is cold-rolled steel containing about 0.4 percent carbon. Shafts are either cold-rolled or forged in sizes from 75 to 125 mm. For sizes above 125 mm, shafts are forged and machined to the required size. Plastic shafts are widely used for light load applications. One advantage of using plastic is safety in electrical applications, since plastic is a poor conductor of electricity.

Components such as gears and pulleys are mounted on shafts by means of key. The design of the key and the corresponding keyway (see Fig. 4.4) in the shaft must be properly evaluated. For example, stress concentrations occur in shafts due to keyways, and the material removed to form the keyway further weakens the shaft.

In addition to satisfying strength requirements, shafts must be designed so that deflections are within acceptable limits. Excessive lateral shaft deflection can hamper gear performance and cause objectionable noise. The associated angular deflection can be very

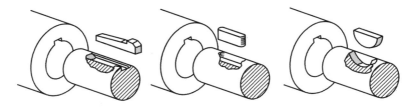

Figure 4.4   Keys and keyways

destructive to non-self-aligning bearings. Self-aligning bearings (Fig. 4.5) may eliminate this trouble if the deflection is otherwise acceptable.

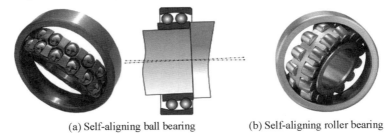

(a) Self-aligning ball bearing   (b) Self-aligning roller bearing

Figure 4.5   Self-aligning bearings

If shafts are run at critical speeds, severe vibrations can occur which can seriously damage a machine. It is important to know the magnitude of these critical speeds so that they can be avoided. As a general rule of thumb, the difference between the operating speed and the critical speed should be at least 20 percent.

Many shafts are supported by three or more bearings, which means that the problem is statically indeterminate. Textbooks on strength of materials give methods of solving such problems. The design effort should be in keeping with the economics of a given situation. For example, if one line shaft supported by three or more bearings is needed, it probably would be cheaper to make conservative assumptions as to moments and design it as though it were determinate. The extra cost of an oversize shaft may be less than the extra cost of an elaborate design analysis.

When axial movement between the shaft and hub is required, relative rotation is prevented by means of splines machined on the shaft and into the hub (see Fig. 4.6). There are two forms of splines: straight-sided splines and involute splines. The former is relatively simple and employed in some machine tools. The latter has an involute curve in its outline, which is in widespread use on gears. The involute form is preferred because it provides for self-centering of the mating element and because it can be machined with standard hobs used to cut gear teeth.

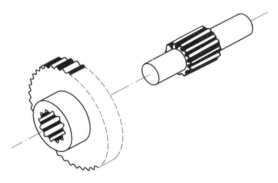

Figure 4.6 General form of spline connection

Words and Expressions

shaft [ʃɑːft] n. 轴,
spline [splain] n. 花键; vt. 用花键联接
rectangular [rekˈtæɡjulə] a. 矩形的
slotted screw driver blade 一字螺丝刀头
torsional [ˈtɔːʃənəl] a. 扭转的
torque [tɔːk] n. 扭矩,转矩

mounted ['mauntid] *a.* 安装好的,固定好的
bend [bend] *v.* ;*n.* 弯曲
bending deflection 弯曲变形,弯曲挠度
reversal [ri'və:səl] *n.* 颠倒,反转,反向
cold-rolled 冷轧的,冷态滚压的
forge [fɔ:dʒ] *n.* 锻造,锻工车间
machined 经过机械加工的
key[ki:] *n.* 键
keyway ['ki:ˌwei] *n.* 键槽,也可以写为 key seat
self-aligning bearing 调心轴承(一滚道是球面形的,能对两滚道轴心线间的角偏差及角运动作适应性自调整的滚动轴承)
critical speed 临界转速,临界速度
rule of thumb 单凭经验的方法,经验法则,靠经验估计
operating speed 工作转速,工作速度
indeterminate [ˌindi'tə:minit] *a.* 不确的,不确定的
statically indeterminate 静不定的,超静定的
conservative [kən'sə:vətiv] *a.* 保守的
straight-sided spline 矩形花键
involute ['invəluːt] *n.* 渐开线
self-centering 自动定心的
mating element 配合零件
hob [hɔb] *n.* 滚刀
general form 常见形式,通用形式

5. Threaded Fasteners

Threaded fasteners perform the function of locating, clamping, adjusting, and transmitting force from one machine member to another. They are thoroughly standardized and generally designed for use in mass production of machines. The use of threaded fasteners (see Fig. 5.1) remains the basic assembly method in the design and construction of machines despite advances in other methods of joining. Thus the designers must select standard fasteners of the type and size that will most adequately suit the application at hand.

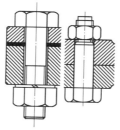

Figure 5.1 Threaded fasteners

A threaded fastener is a device that can effectively exert and maintain a large force in one direction (axially) through the application of a small force in another direction (tangentially). All are based on the single-threaded screw, a simple machine that yields a large mechanical advantage in minimal space and theoretically is self-locking. Effective use, however, requires the aid of two other simple machines: the lever and the wheel-and-axle machine. The wrench is basically a lever; the screwdriver is a wheel-and-axle machine.

Threaded fasteners are basically small, highly stressed tensile components. Threads are helical ridges formed by cutting or cold forming a groove onto the surface of a cylindrical bar, thus producing what is known as a screw, bolt, or stud. Threads are also formed internally in cylindrical holes and constitute what is known as a nut. Matching external and internal threads so that they may be assembled is the key to all threaded fasteners. The rotary motion of a nut against a stationary screw first imparts an axial movement along the screw. When resistance is encountered, the threads generate an axial force. Further rotation demands increased effort (torque), with a resulting increase in axial force. Thus the connection remains tight unless some external influence such as vibration or temperature change overrides the initial condition.

Unified standard thread is available in three basic series of diameter-pitch combinations. The coarse-thread series (UNC) is most common and is recommended for general assembly use where vibration is not a problem and where frequent disassembly is required. The fine-thread series (UNF) is somewhat stronger in tension and is more suitable where fine adjustment may be required. This series is often used in aircraft assemblies. The extra-fine-thread series (UNEF) is employed where the mating external thread is in a thin-walled member. This series is also more resistant to vibration and provides for very fine adjustments.

Unified standard threads are identified on drawings, parts lists, and so forth, with a shorthand notation that includes size, thread series, class of fit, external or internal thread symbol (A is for external threads, B is for internal threads), and the hand of the thread. For example, the designation

$$\frac{1}{4}-20\text{UNC}-2\text{A}-\text{RH}$$

identifies an externally threaded part having a basic major diameter of $\frac{1}{4}$ in., unified coarse thread with 20 threads per inch, tolerance class 2, and right-hand threads. Usually the hand designation is omitted for right-hand, since these are standard. Another example is

$$\frac{3}{4}-16\text{UNF}-2\text{B}-\text{LH}$$

which designates an internal, left-hand thread having a basic major diameter of $\frac{3}{4}$ in., and 16 unified fine threads per inch of length.

ISO standards include many diameter-pitch combinations, but proposed U.S. standards specify a single diameter pitch series. Metric threads are designated by the letter M followed by basic major diameter in millimeters, which is then followed by the pitch in millimeters separated by the symbol " ×. " For example, M4 × 0.7 specifies a metric thread with basic major diameter of 4 mm and pitch of 0.7 mm.

Washers are often used with threaded fasteners to provide a better bearing surface for nuts and bolt heads, to provide a bearing surface over large clearance holes or slots, to distribute the load over a larger area, to prevent marring of parts during assembly, to improve torque-tension ratio (by reducing friction), and to provide locking, in some cases through spring action. Flat washers are thin, annular-shaped disks used primarily for bearing surface and load distribution and have no locking capability. Helical spring lockwashers are essentially single-coil helical springs that flatten under load. Spring action assists in maintaining the bolt load, while split edges provide some locking action by biting into the bearing surfaces. Lock washers are generally made of hardened steel.

Carbon steel and steel alloys are most commonly used fastener materials. Many stainless steel alloys are used for threaded fasteners where improved corrosion resistance is required. Nickel-based superalloys such as Monel and Inconel are used for fasteners where strength at high temperatures as well as corrosion resistance is required. Aluminum, bronze, and brass are also used for threaded fasteners. In corrosive environments where great strength is not required, fasteners made of nylon and other plastics are both suitable and economical.

Words and Expressions

threaded fastener 螺纹紧固件,螺纹连接件
design for 为…而设计
mass production 大规模生产
joining ['dʒɔiniŋ] n. 连接,连接物
tangentially [tæn'dʒənʃəli] ad. 成切线地
single-threaded screw 单头螺纹螺钉
mechanical advantage 机械效益,机械增益
self-locking [self'lɔkiŋ] a. 自锁的
wheel-and-axle 轮轴
ridge [ridʒ] n. 隆起物,凸起
cold forming 冷成型,冷态成型
stud [stʌd] n. 双头螺柱(两头均有螺纹的圆柱形紧固件)
impart [im'pɑːt] v. 给予
wedge [wedʒ] n. 楔,楔形物
unified standard 统一标准
override [ˌəuvə'raid] v. 超过,克服
pitch [pitʃ] n. 螺距
diameter-pitch combination 直径—螺距组合
UNC (unified coarse thread) 统一标准粗牙螺纹
UNF (unified fine thread) 统一标准细牙螺纹
UNEF (unified extra-fine thread) 统一标准超细牙螺纹
fine adjustment 精密调整
part list 零件表,明细表,材料清单
and so forth 等等
shorthand ['ʃɔːthænd] n. 简略的表示方法
shorthand notation 简化符号
class of fit 配合级别

right hand thread 右旋螺纹
basic major diameter 外径的基本尺寸,公称直径
major diameter (螺纹)大径,外径
metric thread 公制螺纹
screw fastener 螺丝紧固件
washer ['wɔʃə] n. 垫圈
bearing surface 支承面
mar [mɑ:] v. 损坏,破坏
annular ['ænjulə] a. 环形的,环的
lockwasher 防松垫圈
helical spring lockwasher 弹簧垫圈
single-coil helical spring 单圈螺旋形弹簧
harden [hɑ:dən] n. 淬火,淬硬
nickel ['nikl] n. 镍
superalloy [sju:pə'ælɔi] n. 超耐热合金,高温合金
Monel [məu'nel] n. 蒙乃尔铜镍合金
Inconel 因科合金,铬镍铁耐热耐腐蚀合金

6. Lubrication

Although one of the main purposes of lubrication is to reduce friction, any substance—liquid, solid, or gaseous—capable of controlling friction and wear between sliding surfaces can be classed as a lubricant.

Varieties of lubrication

Unlubricated sliding. Metals that have been carefully treated to remove all foreign materials seize and weld to one another when slid together. In the absence of such a high degree of cleanliness, adsorbed gases, water vapour, oxides, and contaminants reduce friction and the tendency to seize but usually result in severe wear; this is called "unlubricated" or dry sliding.

Fluid-film lubrication. Interposing a fluid film that completely separates the sliding surfaces results in fluid-film lubrication. The fluid may be introduced intentionally as the oil in the main bearings of an automobile, or unintentionally, as in the case of water between a smooth rubber tire and a wet pavement. Although the fluid is usually a liquid such as oil, water, and a wide range of other materials, it may also be a gas. The gas most commonly employed is air.

To keep the parts separated, iu is necessary that the pressure within the lubricating film balance the load on the sliding surfaces. If the lubricating film's pressure is supplied by an external source, the system is said to be lubricated hydrostatically. If the pressure between the surfaces is generated as a result of the shape and motion of the surfaces themselves, however, the system is hydrodynamically lubricated. This second type of lubrication depends upon the viscous properties of the lubricant.

Boundary lubrication. A condition that lies between unlubricated sliding and fluid-film lubrication is referred to as boundary lubrication, also defined as that condition of lubrication in which the friction between surfaces is determined by the properties of the surfaces and properties of the lubricant other than viscosity. Boundary lubrication encompasses a significant portion of lubrication phenomena and commonly occurs during the starting and stopping of machines.

Solid lubrication. Solids such as graphite and molybdenum disulfide are widely used when normal lubricants do not possess sufficient resistance to load or temperature extremes. But lubricants need not take only such familiar forms as fats, powders, and gases;

even some metals commonly serve as sliding surfaces in some sophisticated machines.

Functions of lubricants

Although a lubricant primarily controls friction and wear, it can and ordinarily does perform numerous other functions, which vary with the application and usually are interrelated.

Friction control. The amount and character of the lubricant made available to sliding surfaces have a profound effect upon the friction that is encountered. For example, disregarding such related factors as heat and wear but considering friction alone between two oil-film lubricated surfaces, the friction can be 200 times less than that between the same surfaces with no lubricant. Under fluid-film conditions, friction is directly proportional to the viscosity of the fluid. Some lubricants, such as petroleum derivatives, are available in a great range of viscosities and thus can satisfy a broad spectrum of functional requirements. Under boundary lubrication conditions, the effect of viscosity on friction becomes less significant than the chemical nature of the lubricant.

Wear control. Wear occurs on lubricated surfaces by abrasion, corrosion, and solid-to-solid contact. Proper lubricants will help combat each type. They reduce abrasive and solid-to-solid contact wear by providing a film that increases the distance between the sliding surfaces, thereby lessening the damage by abrasive contaminants and surface asperities.

Temperature control. Lubricants assist in controlling temperature by reducing friction and carrying off the heat that is generated. Effectiveness depends upon the amount of lubricant supplied, the ambient temperature, and the provision for external cooling. To a lesser extent, the type of lubricant also affects surface temperature.

Corrosion control. The role of a lubricant in controlling corrosion of the surfaces themselves is twofold. When machinery is idle, the lubricant acts as a preservative. When machinery is in use, the lubricant controls corrosion by coating lubricated parts with a protective film that may contain additives to neutralize corrosive materials. The ability of a lubricant to control corrosion is directly related to the thickness of the lubricant film remaining on the metal surfaces and the chemical composition of the lubricant.

Other functions

Lubricants are frequently used for purposes other than the reduction of friction. Some of these applications are described below.

Power transmission. Lubricants are widely employed as hydraulic fluids in fluid transmission devices.

Insulation. In specialized applications such as transformers and switchgear, lubricants with high dielectric constants act as electrical insulators. For maximum insulating properties, a lubricant must be kept free of contaminants and water.

Shock dampening. Lubricants act as shock-dampening fluids in energy-transferring devices such as shock absorbers and around machine parts such as gears that are subjected to high intermittent loads.

Sealing. Lubricating grease frequently performs the special function of forming a seal to retain lubricants or to exclude contaminants.

Words and Expressions

lubrication [luːbriˈkeiʃən] *n.* 润滑
friction [ˈfrikʃən] *n.* 摩擦,摩擦力

wear [wɛə] v. ;n. 磨损,损耗,磨蚀
sliding ['slaidiŋ] n. ;a. 滑动,可相互移动
lubricant ['lju:brikənt] n. 润滑剂,润滑材料
unlubricated [ʌn'lu:brikeitid] a. 无润滑的
foreign material 外来材料,异物,杂质
seize [si:z] v. (机器等)卡住,咬住,粘结
weld [weld] n. ;v. 焊接,熔接
cleanliness ['klenlinis] n. 清洁度,洁净
adsorb [æd'sɔ:b] v. 吸附,吸取
contaminant [kən'tæminənt] n. 污染物,杂质
intentionally [in'tenʃənli] ad. 故意地
hydrostatical [haidrəu'stætikəl] a. 流体静力(学)的,液压静力的
hydrodynamical [haidrəudai'næmikəl] a. 流体动力(学)的
viscous ['viskəs] a. 粘的,粘性的,粘稠的
boundary lubrication 边界润滑
viscosity [vis'kɔsiti] n. 粘性,粘滞度
encompass [in'kʌmpəs] v. 环绕,包围,包括,包含
graphite ['græfait] n. 石墨
molybdenum [mɔ'libdinəm] n. 钼
disulfide [dai'sʌlfaid] n. 二硫化物
temperature extremes 温度极限
fat [fæt] n. 脂肪,油脂;a. 油脂的,多脂的
profound [prə'faund] a. 深奥的,深刻的,极度的
derivative [di'rivətiv] n. 衍生物;a. 派生的,衍生的
spectrum ['spektrəm] n. 光谱,领域,范围,系列
abrasion [ə'breiʒən] n. 擦伤,磨损,磨耗
corrosion [kə'rəuʒən] n. 腐蚀,侵蚀,锈蚀
lessen ['lesn] v. 减少,缩小,减轻
asperity [æs'periti] n. 粗糙,凹凸不平
ambient ['æmbiənt] a. 周围的;n. 周围环境

provision [prə'viʒən] n. （预防）措施,保证,保障
to a lesser extent 在较小的程度上
preservative [pri'zə:vətiv] a. 保存的,防腐的;n. 防腐剂,保存剂
additive ['æditiv] n. 添加剂,外加物
neutralize ['nju:trəlaiz] v. 使中和,使中立
insulation [insju'leiʃən] n. 绝缘,隔热,绝缘体
transformer [træns'fɔ:mə] n. 变压器
switchgear ['switʃgiə] n. 开关装置,配电装置
dielectric [daii'lektrik] a. 不导电的,绝缘的,介电的
dielectric constant 介电常数,介质常数
dampen ['dæmpən] v. 抑制,使衰减,阻尼,减震,缓冲
absorber [əb'sɔ:bə] n. 减震器,缓冲器,阻尼器
intermittent [intə'mitənt] a. 间歇的,断续的,周期性的
sealing ['si:liŋ] n. 密封,封接,封口
grease [gri:s] n. 润滑脂,黄油
retain [ri'tein] v. 保留,保持不变,留住
exclude [iks'klu:d] v. 拒绝,排除,隔绝

7. Machine Tool Frames

The frame is a machine's fundamental element. Most frames are made from cast iron, welded steel, composite, or concrete. The following factors govern material choice.

The material must resist deformation and fracture. Hardness must be balanced against elasticity. The frame must withstand impact, yet yield under load without cracking or permanently deforming. The frame material must eliminate or block vibration transmission to reduce oscillations that degrade accuracy and tool life. It must withstand the hostile shop-floor environment, including the newer coolants and lubricants. Material expansion must be understood to minimize forces needed to move slides. The material must not build up too much heat, must retain its shape for its lifetime, and must be dense enough to distribute forces throughout the machine.

Pros and Cons

Either castings or welded sections can be used in most applications. The decision on which is best depends on the costs in a given design situation.

Cast iron. Almost all machine tool frames were traditionally made of cast iron because features difficult to obtain any other way can be cast in. Castings have a good stiffness-to-weight ratio and good damping qualities. Modifying wall thickness and putting the metal where it's needed is fairly easy.

Although cast iron is a fairly cheap material, each casting requires a pattern. Larger sizes are a limiting factor because of pattern cost, problems with bolted joints, and the need to anneal castings, which is difficult and costly with larger sections. Smaller, high-volume machines usually have cast iron frames because they more easily absorb pattern cost. Welded frames may be cheaper for lower volume machines.

Welded steel. Machine builders fabricate steel frames from welded steel sections when casting is impractical. Because steel has a higher modulus, it is usually ribbed to provide stiffness. The number of welds is a design tradeoff: with welding, it's easy to make large sections and add features even after the initial design is complete, but the heat can introduce distortion and also adds cost. Welds also help block vibration transmission through the steel frame. Builders sometimes increase damping by circulating coolant through the welded structure or adding lead or sand to frame cavities.

Composites. Advanced forms of these materials, including those with polymer, metal, and ceramic matrices, may change machine tool design dramatically. Both matrix and reinforcing material can be tailored to provide strength in specific axes.

Designers must consider the different expansion coefficients between the composite and the metal sections to which it is joined. The most common applications for this material are high-accuracy machine tools and grinders.

Foundations

Foundations ensure the machine's stiffness; shock absorption and isolation are secondary considerations. In selecting a foundation, designers must consider the machine's weight, the forces it generates, accuracy requirements, and the loads being transmitted to the ground by adjacent machines. Soil condition can be a problem because long-term changes can influence machine stability.

Frame Design

The major considerations in frame design are loads, damping, apertures, heat distortion, and noise.

Loads. Understanding the static and dynamic loads a machine generates is essential. The basic load is static: the mass of the machine and its workpiece. The dynamic load adds all that happens once the machine is running. This includes the forces of acceleration, deceleration, tool action, irregular loads caused by an unbalanced condition, or self-exciting loads from load and vibration interaction. Finite element analysis (Fig. 7.1) gives a good indication of how a frame will react to loads.

Damping. Though frame material and design should handle damping, dampers are sometimes built into frame sections to handle specific problems. They are effective only when the designer has a

MACHINE COMPONENTS AND DESIGN

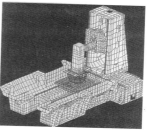

Figure 7.1 A machine tool frame and a FEA model

good understanding of all the loads involved. For example, a damper that works well under static conditions may do more harm than good under dynamic conditions.

Apertures. Each frame face should be solid, but the machine needs openings for assembly and maintenance. The designer balances aperture number and size against stiffness and strength requirements.

Thermal considerations. Heat from external or internal sources can be a major cause of error if the frame distorts. External sources include ambient shop conditions, cooling and lubricating media, and the sun. The machine also has its own heat sources: motors, friction from machine motion, and the cutting action of the tool on the workpiece. Ideally, frame heating should be minimized and kept constant.

Noise. Reduction of noise for health and safety reasons is a fairly recent concern. Air turbulence from moving parts and fans can be a particular problem. Enclosures prevent sound transfer through the machine, and sound damping materials help reduce objectionable sound.

Words and Expressions

machine tool frame 机床机架

composite [ˈkɔmpəzit] a. 合成的,复合的;n. 复合材料,混合料
deformation [diːfɔːˈmeiʃən] n. 变形
fracture [ˈfræktʃə] v. ;n. 破碎,断裂
elasticity [elæsˈtisiti] n. 弹性(变形),伸缩性,弹性力学
oscillation [ɔsiˈleiʃən] n. 振荡,摆动
accuracy [ˈækjurəsi] n. 准确,精度,精确性
shop-floor 车间,工场
coolant [ˈkuːlənt] n. 冷却液,切削液,乳化液
slide [slaid] n. 滑板(可沿导轨移动的部件),滑块
pros and cons 正反面,优缺点,正反两方面的理由
casting [ˈkɑːstiŋ] n. 浇铸,铸件
stiffness [ˈstifnis] n. 刚性,刚度
pattern [ˈpætən] n. 模,木模,铸模
bolt [bəult] n. 螺栓;v. 用螺栓固定
joint [dʒɔint] n. ;v. 结合,连接
anneal [əˈniːl] v. ;n. (使)退火,(加热后)缓冷
absorb [əbˈsɔːb] v. 吸收,吸引,承担费用
modulus [ˈmɔdjuləs] (pl. moduli) n. 模数,模量,系数
rib [rib] n. 肋,筋;v. 加肋,加筋
tradeoff = trade-off 折衷(方案,方法),权衡
distortion [disˈtɔːʃən] n. 变形,扭曲,畸变
lead [led] n. 铅,铅制品
cavity [ˈkæviti] n. 空腔,(铸造)型腔,室穴,腔体,凹处
polymer [ˈpɔlimə] n. 聚合物,高分子材料
ceramics [siˈræmiks] n. 陶瓷
matrix [ˈmeitriks] (pl. matrices 或 matrixes) n. 基体,本体
tailor...to... 使…适合(满足)…(的要求,需要,条件)
stability [stəˈbiliti] n. 平衡状态,稳定性,耐久性
brittle [ˈbritl] a. 易碎的,脆的
conductivity [kɔndʌkˈtiviti] n. 传导率,传导系数

damp [dæmp] n. ;v. 阻尼,衰减
stress relieving 应力消除热处理
accommodate [ə'kɔmədeit] v. 供应,适应,向…提供,容纳
insert 嵌入件,埋入件
expansion coefficient 膨胀系数
foundation [faun'deiʃən] n. 基础,地基,底座,机座
mounting ['mauntiŋ] n. 安装,安置,装配,固定件
aperture ['æpətjuə] n. 孔,壁孔,孔隙,孔径
deceleration [di:selə'reiʃən] n. 减速(度),降速
self-exciting 自激的,自励的
FEA 有限元分析
assembly [ə'sembli] n. 装配
maintenance ['meintinəs] n. 维持,(技术)保养,维修
thermal ['θə:məl] a. 热量的,热力的,热的
workpiece ['wə:kpi:s] n. 工件
turbulence ['tə:bjuləns] n. 湍流,旋涡
enclosure [in'kləuʒə] n. 外壳,机壳,罩
objectionable [əb'dʒektʃənəbl] a. 引起反对的,不适合的,有害的

8. Engineering Graphics

Engineering graphics is a cornerstone of engineering. The essence of engineering—that is, design—requires graphics as the means of communication within the design process.

Study of the fundamentals of engineering graphics is one key to your success as an engineer. Being able to describe an idea with a sketch is a prerequisite of the engineering profession. The ability to put forth a three-dimensional geometry in a form that can be communicated to other engineers, scientists, technicians, and nontechnical personnel is a valuable asset. Of equal importance is knowing how to read and understand the graphics prepared by others.

The ability to communicate is the key to success for a practicing engineer. Graphic communication, along with written and oral communication, constitutes an important part of a program of study in engineering. The fundamentals of the graphics language are universal in the industrialized world, an advantage not afforded by the written and spoken language. Thus graphics may be said to be "a language for engineers."

The study of graphics involves three aspects: terminology, skills and theory. Definitions of general terms encountered in graphic applications are introduced below.

Engineering graphics is the area of engineering which involves the application of graphic principles in the development and conveyance of design concepts.

Engineering design is the systematic process by which a solution to a problem is created. Engineering graphics provides visual support, a basis for engineering analysis, and documentation for the design process.

Descriptive geometry is a set of principles which enable the geometry of an object to be identified and delineated by graphic means. It is the theory by which spatial (three-dimensional) problems involving angles, shapes, sizes, clearances, and intersections are solved with two-dimensional representation.

Computer graphics utilizes the digital computer to define, manipulate, and display devices, processes, and systems for the purpose of analysis, design, and communication of engineering solutions.

Geometric modeling is the representation of a concept, process, or system operation usually in a mathematical form, and more specifically as an electronic database. Computer-based geometric modeling may conveniently be classified as wireframe, surface, or solid.

Engineering graphics is in a period of rapidly changing graphics technology. The traditional tools of graphics, such as the T-square, compass, and drafting machines, are being displaced by computer hardware and software. We are in an exciting era in which we will experience the transition from scales, triangles, and dividers to a computer keyboard and from blueprints to databases.

The engineers of today see the engineering drawing as a by-product of the CAD process. The control of the design-manufacture cycle is now the electronic database of the design. Changes are incorporated instantly in all aspects of the design. New product models can be quickly developed and oftentimes proved with computer simulations, thus bypassing the prototype. If drawings are desired for manufacture or documentation, they may be quickly obtained from the database.

The engineering student of today will study graphics from the standpoint of supporting the design process. Geometric modeling techniques, analysis techniques which are mathematically based, and practice in visualization of three-dimensional geometries will be the focus of intensive computer utilization. In order to prepare concepts for modeling and analysis, freehand techniques will be studied and practiced. The student will learn to produce and interpret multiviews and pictorials both via sketches and computer techniques. Many of the graphics standards for appropriate representation of object features (sections, dimensioning, multiviews) will be studied.

Working in three dimensions with the computer, graphics will be produced easily in two-or three-dimensional modes depending upon the application. Creating two-dimensional graphics such as xy plots and schematics will be accomplished with CAD software. Two-dimensional geometric primitives such as circles and rectangles which serve as the generating geometry for cylinders and prisms are a part of two-dimensional software. Special applications, for example, dimensioning, generally utilize a two-dimensional view or series of two-dimensional views.

Three-dimensional geometric modeling involves wireframe, surface, and solid models. A wireframe model shows a series of nodes connected by lines to form an object. A surface model completely defines a series of areas connected to form the boundary of an object. The solid model is a total definition of an object which includes knowledge of all boundary and internal points. From a solid model, a complete analysis of performance of the object can be performed on the computer with appropriate software.

Computer graphics has become a powerful design tool which promises to enhance significantly the engineer's ability to be creative and innovative in the solution of complex problems.

The information revolution is well under way. The rapid advancement of electronic technology has changed the way we work and live. The language of graphics will continue to be a cornerstone of communication for engineers and other technical persons. However, the changes we are seeing in the methods of transferring graphic, written, and spoken material is astounding. These changes are improving the productivity of industries and individuals as well as increasing the quality of products and the working environment. The requirements for the twenty-first century engineer will include a sound understanding of the fundamentals of graphics and the implementation of graphics to support the design process.

Words and Expressions

graphics ['græfiks] *n.* 制图学,图形学,图解
engineering graphics 工程图学
cornerstone ['kɔːnəstəun] *n.* 基础,基石
essence ['esns] *n.* 本质,要素,精华
means of communication 交流工具(手段),通讯工具(手段)
fundamental [fʌndə'mənt] *n.* 基本原理;*a.* 基本的
prerequisite [priː'rekwizit] *n.* 先决条件 *a.* 首要必备的
put forth 提出,发表,拿出
nontechnical [nɔn'teknikl] *a.* 非技术性的
practicing ['præktisiŋ] *a.* 开业的,从业的,在工作的
industrialized [in'dʌstriəlɑizd] *a.* 工业化的
conveyance [kən'veiəns] *n.* 运送,传送
descriptive geometry 画法几何
delineate [di'lineit] *v.* 描绘,描述
geometric modeling 几何建模,几何模型建立,形状模型化
T-square 丁字尺
compass ['kʌmpəs] *n.* 圆规
drafting ['drɑːftiŋ] *n.* 制图,起草
scale 比例尺,刻度尺
divider [di'vaidə] *n.* 分规,两脚规
blueprint ['bluːprint] *n.* 蓝图,设计图
database ['deitbeis] *n.* 数据库
by-product ['bai-prɔdəkt] *n.* 副产品,副产物
prototype ['prəutətaip] *n.* 原型,样机,样品
visualization [vizjuəlai'zeiʃən] *n.* 使看得见的
freehand ['friːhænd] *a.* 徒手画的
pictorial [pik'tɔːriəl] *a.* 用图表示的,图解的

section 截面
dimensioning 标注尺寸
plot [plɔt] n. 曲线,图形
schematic [ski'mætik] a. 示意的,简略的;n. 简图
node [nəud] n. 节点,交点,中心点
primitive ['primitiv] a. 原始的,基本的;n. 基元,图元
geometric primitive 几何图元
rectangle ['rektæŋgəl] n. 长方形,矩形
prism ['prizm] n. 棱柱
wireframe model 线框模型
surface model 曲面模型
solid model 实体模型
innovative ['inəuveitiv] a. 创新的,革新的
astound [əs'taund] v. 使…大吃一惊,令人震惊
sound 可靠的,合理的;彻底地,充分地

9. Sectional Views

Sectional views, frequently called sections, are used to show internal construction of an object that is too complicated to be shown clearly by regular views containing many hidden lines. To produce a sectional view, a cutting plane is assumed to be passed through the part, and then removed.

Section Lining. Section lining can serve a double purpose. It indicates the surface that has been cut and makes it stand out clearly, thus helping the observer to understand the shape of the object. Section lining may also indicate the material from which the object is made. The symbols used to distinguish between different materials in sections are shown in Fig. 9.1. However, there are so many different materials used in design that the general symbol (i.e., the one used for cast iron) may be used for most purposes on technical drawings. The actual type of material required is then noted in the title block or parts list, or entered as a note on the drawing.

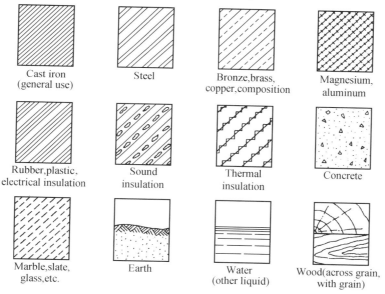

Figure 9.1 ANSI standard section lining symbols for various materials

Full Sections. When the cutting plane extends entirely through the object in a straight line and the front half of the object is imagined to be removed, a full section is obtained. This type of section is used for both detail and assembly drawings. When the section is on an axis of symmetry, it is not necessary to indicate its location. Figure 9.2 shows a full section with cutting plane omitted.

Half Sections. If a cutting plane passes half way through an object, the result is a half section. The half section is used when you wish to show both internal construction and the exterior view of an object in the same view. This type of section is most effective when the object is symmetric or nearly so. Therefore, one half of the object illustrates internal construction, and the other half shows an external view with hidden lines omitted. The half section is visualized by imagining that one quarter of the object has been removed. Figure 9.3 shows the position of the cutting plane in the top view and the resulting half section in the front view. The section view and the external view are separated by a centerline. The hidden lines in the external view portion have been omitted according to conventional practice, although they could be added if needed to clarify the drawing.

The half section is not widely used in detail drawings because of difficulties in dimensioning internal shapes that are shown in part only in the sectioned half (Fig. 9.3). The greatest usefulness of the half section is in assembly drawing, in which it is often necessary to show both internal and external constructions on the same view.

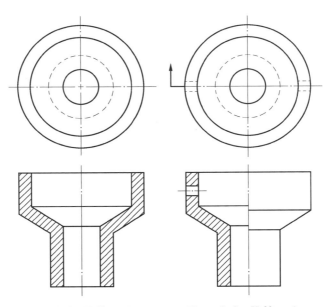

Figure 9.2 Full section   Figure 9.3 Half section

Broken-Out Sections. Where a sectional view of only a portion of the object is needed, broken-out sections may be used (Fig. 9.4). An irregular break line is used to show the extent of the section.

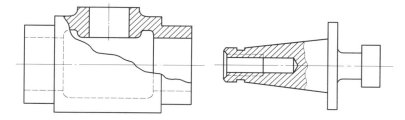

Figure 9.4   Broken-out sections

Offset Sections. In order to include features that are not in a straight line, the cutting plane may be offset, so as to include several planes. Such a section is called an offset section. Figure 9.5 shows a typical use of an offset section.

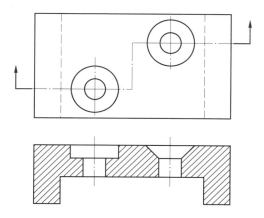

Figure 9.5   Offset section

Aligned Sections. To include in a section certain angled features, the cutting plane may be bent to pass through those features. The features cut by the cutting plane are then imagined to be revolved into a plane. Figure 9.6 is an example of aligned section.

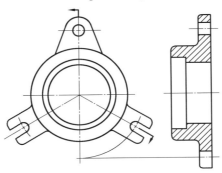

Figure 9.6 Aligned section

Placement of Sectional Views. Whenever practical, sectional views should be projected perpendicularly to the cutting plane and be placed in the normal position for third-angle projection. When the preferred placement is not practical, the sectional view may be removed to some other convenient position on the drawing, but it must be clearly identified, usually by two identical capital letters, one at each end of the line. Normally, the sections are labeled alphabetically, beginning with A-A, then B-B, and so on (Fig. 9.7). Note that view C-C is not a sectional view since the cutting plane is external to the object.

Parts not Section-Lined. General purpose section lining is recommended for most assembly drawings. The section line should be drawn at an angle of 45° with the main outline of the view. On adjacent parts, the section lines should be drawn in the opposite direction, as shown in Fig. 9.8.

Shafts, bolts, nuts, screws, rivets, pins, washers, and gear teeth, should not be section-lined even though the cutting plane passes through them, except that a broken-out section of the shaft may be used to describe more clearly the key, key seat, or pin (Fig. 9.8).

MACHINE COMPONENTS AND DESIGN 45

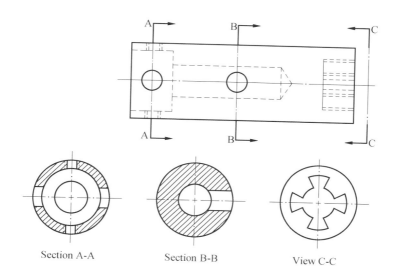

Figure 9.7　Detail drawing having two sectional views

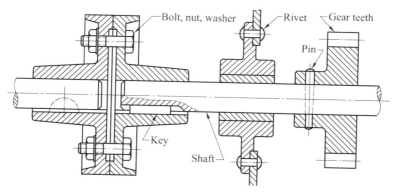

Figure 9.8　Parts that are not section-lined

Words and Expressions

sectional view 剖视图,剖面图,也可以写为 section view
view 视图,外形图

section n.;v. 剖视,剖面,剖面图,断面
internal construction 内部结构
hidden line 虚线
section lining 剖面线
section lining symbol 剖面符号
serve a double purpose 起双重作用
stand out clearly 清晰地显示出
distinguish between 区别,分辨
section lining symbol 剖面符号
general symbol 通用符号
for most purposes 在大多数情况下,在大多数场合
general use 通用,一般用途
parts list 零件明细表
marble ['maːbl] n. 大理石
slate [sleit] n. 石板,石片
across grain 横纹,与木纹垂直,横剖面
with grain 顺纹,顺纹理,纵剖面
ANSI American National Standards Institute 美国国家标准协会
full section 全剖视图
axis of symmetry 对称轴,对称轴线
detail drawing 零件图(表示零件结构、大小及技术要求的图样)
assembly drawing 装配图(表示产品及其组成部分的连接、装配关系的图样)
half section 半剖视图
symmetric [si'metrik] a. 对称的,平衡的;n. 对称
top view 俯视图,顶视图(美国、加拿大等国家的工程图样是采用第三角投影绘制的。在第三角投影中,由上向下投影得到的为顶视图,顶视图配置在前视图的上方)
front view 主视图,正视图,前视图(在第三角投影中,由前向后投影得到的为前视图)

centerline ['sentəlain] 中心线
external view 外观图, 外形视图
dimension [di'menʃən] v. 在……上标出尺寸
broken-out section 局部剖视图, 也可以写为 partial section
break line 波浪线
offset ['ɔ:fset] a. 偏移的, 横向移动的
offset section 转折剖(相当于原 GB 中的阶梯剖), 用几个相互平行的剖切平面, 剖开物体后获得的剖视图, 几个平行剖切平面获得的剖视图
a typical use of 一种典型的用法
aligned section 摆正剖(相当于原 GB 中的旋转剖), 用几个相交的剖切平面, 剖开物体后获得的剖视图, 几个相交剖切平面获得的剖视图
other than M 除了 M, M 除外, 与 M 不同的
label ['leibl] n. 标签, 标志; v. 标注
alphabetically [ˌælfə'betikəli] ad. 按字母(表)顺序, 按 ABC 顺序
placement of sectional view 剖视图的配置
project ['prɔdʒekt] n.; v. 投射, 投影
third-angle projection 第三角投影
main outline 主要轮廓线
section line n.; v. 剖面线, 画剖面线
key seat 键槽, 也可以写为 keyway

10. Engineering Design

Engineering design is a systematic process by which solutions to the needs of humankind are obtained. The process is applied to problems (needs) of varying complexity. For example, mechanical engineers will use the design process to find an effective, efficient method to convert reciprocating motion to circular motion for the drivetrain in an internal combustion engine; electrical engineers will use the process to design electrical generating systems using falling water as the power source; and materials engineers use the process to design ablative materials which enable astronauts to safely reenter the earth's atmosphere.

The vast majority of complex problems in today's high-technology society depend for solution not on a single engineering discipline, but on teams of engineers, scientists, environmentalists, economists, and legal personnel. Solutions are not only dependent upon the appropriate applications of technology but also upon public sentiment as executed through government regulations and political influence. As engineers we are empowered with the technical expertise to develop new and improved products and systems, but at the same time we must be increasingly aware of the impact of our actions on society and the environment in general and work conscientiously toward the best solution in view of all relevant factors.

A formal definition of engineering design is found in the curriculum guidelines of the Accreditation Board for Engineering and Technology (ABET). ABET accredits curricula in engineering schools and derives its membership from the various engineering professional societies. Each accredited curriculum has a well-defined design component which falls within the ABET guidelines. The ABET statement on design reads as follows:

Engineering design is the process of devising a system, component, or process to meet desired needs. It is a decision making process (often iterative), in which the basic sciences, mathematics, and engineering sciences are applied to convert resources optimally to meet a stated objective. The engineering design component of a curriculum must include most of the following features: development of student creativity, use of open-ended problems, development and use of modern design theory and methodology, formulation of design problem statements and specifications, consideration of alternative solutions, feasibility considerations, production processes, concurrent engineering design, and detailed system descriptions. Further, it is essential to include a variety of realistic constraints such as economic factors, safety, reliability, aesthetics, ethics, and social impact.

If anything can be said about the last half of the twentieth century, it is that we have had an explosion of information. The amount of data that can be uncovered on most subjects is overwhelming. People in the upper levels of most organizations have assistants who condense most of the things that they must read, hear, or watch. When you begin a search for information, be prepared to scan many of your sources and document their location so that you can find them easily if the data subsequently appear to be important.

Some of the sources that are available include the following:

1. Existing solutions. Much can be learned from the current status of solutions to a specific need if actual products can be located, studied and, in some cases, purchased for detailed analysis. An improved solution or an innovative new solution cannot be found unless the existing solutions are thoroughly understood.

2. Your library. Many universities have courses that teach you how to use your library. Such courses are easy when you compare them with those in chemistry and calculus, but their importance should not be underestimated. There are many sources in the library that can lead you to the information that you are seeking. You may find what you need in an index such as the *Engineering Index*. There are many other indexes that provide specialized information. The nature of your problem will direct which ones may be helpful to you.

3. Professional organizations. The American Society of Mechanical

Engineers (ASME) is a technical society that will be of interest to students majoring in mechanical engineering. Each major in your college is associated with not one but often several such societies.

4. Trade journals. They are published by the hundreds, usually specializing in certain classes of products and services.

Money and economics are part of engineering design and decision making. We live in a society that is based on economics and competition. It is no doubt true that many good ideas never get tried because they are deemed to be economically infeasible. Most of us have been aware of this condition in our daily lives. We started with our parents explaining why we could not have some item that we wanted because it cost too much. Likewise, we will not put some very desirable component into our designs because the value gained will not return enough profit in relation to its cost.

Industry is continually looking for new products of all types. Some are desired because the current product is not competing well in the marketplace. Others are tried simply because it appears that people will buy them. How do manufacturers know that a new product will be popular? They seldom know with certainty. Statistics is an important consideration in market analysis. Most of you will find that probability and statistics are an integral part of your chosen engineering curriculum. The techniques of this area of mathematics allow us to make inferences about how large groups of people will react based on the reactions of a few.

Words and Expressions

systematic [sisti'mætik] *a.* 有系统的,成体系的,有计划的
reciprocating [ri'siprəkeitiŋ] *n.* ;*a.* 往复(的,式),来回的,摆动的
drive train 动力传动系统

ablative ['æblətiv] *a.* 烧蚀的,脱落的;*n.* 烧蚀材料
reenter ['riː'entə] *v.* 重新进入,重返大气层,重新加入
sentiment ['sentimənt] *n.* 感情,情绪,意见,感想
government regulation 政府法规
empower [im'pauə] *v.* 授权给,准许,授予…的权利(资格)
conscientiously [kɔnʃi'enʃəsli] *ad.* 认真地,凭良心办事地,有责任心地,尽责地
in view of 鉴于,考虑到,由…看来
curriculum [kə'rikjuləm] (*pl.* curricula) *n.* 课程,学习计划
guideline 方针,准则,指导方针
accreditation [əˌkredi'teiʃən] *n.* 鉴定合格,任命,认证
Accreditation Board for Engineering and Technology(ABET)
工程技术认证委员会
derive [di'raiv] *v.* 得到,取得,衍生出,引出
iterative ['itərətiv] *a.* 反复的,迭代的
optimally ['ɔptiməli] *ad.* 最佳地,最优地,最恰当的,最适宜地
open-ended *a.* 可扩展的,无终止的,能适应未来发展的,未确定的
methodology [meθə'dɔlədʒi] *n.* 方法(学,论),分类法
formulation [fɔːmju'leiʃən] *n.* 列方程式,列出公式,(有系统的)阐述
specification [spesifi'keiʃən] *n.* 规格,规范,技术要求,说明书,详细说明
reliability [rilaiə'biliti] *n.* 可靠性,安全性
aesthetics [iːs'θetiks] *n.* 美学
impact ['impækt] *n.* 碰撞,影响,效果,反响
explosion [iks'plouʒən] *n.* 爆炸,蓬勃发展,激增
overwhelming [əuvə'hwelmiŋ] *a.* 压倒的,不可抵抗的,优势的
condense [kən'dens] *v.* 浓缩,精简,压缩,简要叙述
scan [skæn] *v.* 浏览
innovative ['inəuveitiv] *a.* 创新的,革新的
calculus ['kælkjuləs] *n.* 计算,微积分
underestimate [ʌndər'estimeit] *v.* 低估,看轻

index ['indeks] n. 索引,检索,目录
professional [prə'feʃənl] n. 特性,专长,专业,专业人员,内行
trade journal 行业杂志
infeasible [in'fi:zəbl] a. 不能实行的,办不到的,不可能的
marketplace ['mɑ:kit'pleis] n. 市场
statistics [stə'tistiks] n. 统计,统计学,统计数字
probability [prɔbə'biliti] n. 可能性,概率
inference ['infərəns] n. 推论,结论,含意
more specifically 更准确地说,更具体地说

11. Engineering Design and Safety Factors

Any consideration of engineering materials and engineering design of systems utilizing materials must initially deal with Murphy's law for materials systems and the laws of materials applications:

Murphy's law: If any material can fail, it will.

Laws of materials applications

(1) All materials are unstable.

(2) The materials system is only as strong or as stable as its weakest or most unstable component.

Although they are both obvious and incontrovertible, we might elaborate upon these laws, especially the laws of materials applications. All materials are indeed unstable. Even such materials as platinum can be degraded in particular environments. Under stress, all materials respond to that stress. Creep is an example. Given enough time, failure can be calculated to occur in all materials under creep stress. Stress and environment can act in concert to cause failure, for example, stress corrosion cracking. Creep temperature effects and stress-temperature effects can contribute to the degradation of properties and components and system failure.

The design process

The design process usually begins with a specification of a solution. We sometimes allude to a design cycle, but the process may contain a design cycle plus design implementation, which involves actual production based upon the design. The design cycle can involve the original thoughts, sketches, and knowledge that in the specification stage produce engineering drawings. Computer-aided design is now employed to implement a cycle in which various designs or design ideas may be tested or simulated.

Even in the early design cycle, it is useful to have some concept of available materials. Designs that cannot be achieved because materials with requisite properties do not exist are only concepts that will never be a reality unless a material or materials are developed. Prior to a fabrication step, materials specifications must be made depending upon the design requirements. These may have to be altered in the design simulation if the requisite properties cannot be identified in existing materials, or fabrication may have to be delayed until new materials can be designed and fabricated. Materials are the reality in a design.

Figure 11.1 illustrates a simple schematic representation of a design process that includes design implementation. The role of materials in the implementation is of course crucial because they make the design a reality. The design or design implementation diagram shown in Figure 11.1 may not characterize the final product. The first process may produce a prototype that will be tested and improved upon to develop a final product. Generally, a final product works reliably and is economical (or profitable) and safe.

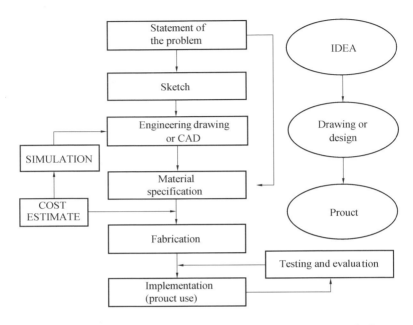

Figure 11.1　Simple schematic diagram showing a design and design implementation process and process cycles

Safety factors in engineering designs

The concept of safety factors is normally applied to the evaluation and consideration of strength of materials used in a design. Safety factors are normally applied to the yield strength of a material but can be applied to the ultimate tensile strength. The yield stress is the stress above which the material (particularly a metal or alloy) will deform permanently. Having exceeded the yield stress or yield strength, materials begin to creep. That is, dislocations will be created, and application of a load to continue the stress above yield will allow the material to slip. Over time, it may slip enough to simply come apart. So if a material is used in a service environment in which the yield stress is never exceeded, it is unlikely it will fail by being over stressed. We sometimes base a design on fractions of the yield stress. For example, if we assume a material will never exceed half the value of the yield stress and use this in the actual design, we can realize a safety factor of 2. In an extreme case, we might allow the maximum stress in the system to be $\sigma_y/4$, and the safety factor would be 4. Obviously a design that would put a material into a service environment in which the normal stresses would be near the ultimate tensile stress for sustained periods of time would be unsafe. On the other hand, large safety factors could render a design unworkable because no materials could be identified with the requisite strengths. In addition, even fulfilling the strength requirements does not necessarily justify a material's use. The material may be too expensive, or too corrosive, or unstable in other ways for the intended application. Or perhaps, in addition to high strength, high electrical conductivity would be required.

The materials system

Finally, let's look at the concept of a materials system. A system in the most general sense involves an arrangement of and connection of parts or elements into a whole. A machine is therefore a mechanical system having connected components. A materials system is therefore an arrangement of connected materials. The system can be partitioned into different materials where they connect, and in the most ideal system materials are separated by an interface or boundary that connects one component to another or provides a transition from one regime to another. Fracture can occur at crystal interfaces, and therefore even a polycrystalline metal or alloy might be thought of as a system of connected crystals.

In the context of failure as we have discussed, it is important to realize that failure cannot only occur within a regime (a component in a connected arrangement), but at the connection itself. So the interface is also a part of the system and must be considered in the context of the weakest link corollary and Murphy's law.

Failure prevention is certainly not a hopeless issue even in very complex materials systems but it does require a great deal of testing, evaluation, and fundamental understanding of materials behavior under a variety of conditions. Selecting optimum conditions, underutilizing components in a system (designing for the weakest link and employing safety factors), introducing safety factors, consideration for environmental changes and even bizarre conditions of use—all contribute to successful applications of products. One need only to take a glimpse at the technology around us to appreciate how successful we have been in designing efficient, dependable, and useful products and fixtures.

Words and Expressions

Murphy's law 墨菲法则(一种幽默的规则,它认为任何可能出错的事终将出错)
unstable [ʌn'steibl] a. 不牢固的,不稳定的
incontrovertible [inkɔntrə'və:təbl] a. 无可争辩的,反驳不了的,颠扑不破的
elaborate [i'læbərit] v. 详细描述,详细阐述
platinum ['plætinəm] n. 铂,白金
creep [kri:p] n. 蠕变,徐变
in concert 一致,共同
allude [ə'lju:d] v. (间接)提到,引证,指…说
implementation [implimen'teiʃən] n. 履行,实施,实现,执行过程
sketch [sketʃ] n. 草图,略图;v. 绘草图
simulation [simju'leiʃən] n. 模拟,仿真
product use 产品用途

MACHINE COMPONENTS AND DESIGN

illustrate ['iləstreit] v. 举例说明,阐明,图解
schematic [ski'mætik] a. 示意性的,图表的;n. 示意图,原理图
schematic representation 图示,略图,简图
crucial ['kru:ʃəl] a. 决定性的,关键的,紧要关系的
diagram ['daiəgræm] n. 图表,示意图,特性曲线
profitable 有利可图的
yield strength 屈服强度
ultimate tensile strength 极限抗拉(拉伸)强度
dislocation [dislə'keiʃən] n. (晶体格子中)位移,位错
apart [ə'pɑ:t] ad. 相隔,相距,分开,拆开
fraction ['frækʃən] n. 分数,比值,几分之一
sustained [səs'teind] a. 持续的,不间断的
render ['rendə] v. 致使
justify ['dʒʌstifai] v. 证明⋯是正确的,认为⋯有理由
partition [pɑ:'titʃən] n. ;v. 划分,区分,分割,分类
interface ['intəfeis] n. 界面,接触面
transition [træn'siʒən] n. 转变,变换,过渡
regime [rei'ʒi:m] n. 方法,领域,范围
polycrystalline [pɔli'kristəlain] a. 多晶的
in the context of 在⋯情况下
the weakest link 最薄弱的环节
corollary [kə'rɔləri] n. 推论,必然的结果
optimum ['ɔptiməm] a. ;n. 最佳(的,值,状态,条件,方式),最适宜,最有利的
underutilize [ʌndə'ju:tilaiz] v. 未充分利用
bizarre [bi'zɑ:] a. 奇怪,奇异的,奇妙的
glimpse [glimps] n. 一看;v. 看见
take a glimpse at 看到
dependable [di'pendəbl] a. 可靠的,可信任的
fixture ['fikstʃə] n. 夹具,设备,装置

12. Computer Applications in Design and Graphics

Computers are widely used in engineering and related fields and their use is expected to grow even more rapidly than in the past. Engineering and technology students must became computer literate, to understand the applications of computers and their advantages. Not to do so will place students at a serious disadvantage in pursuing their careers.

Computer-aided design (CAD) involves solving design problems with the help of computer. In CAD, the traditional tools of graphics, such as the T-square, drawing compass (see Figs. 12.1 and 12.2), and drawing board are replaced by electronic input and output devices. When using a CAD system, the designer can conceptualize the object to be designed more easily on the computer screen and can consider alternative designs or modify a particular design quickly to meet the necessary design requirements or changes. The designer can then subject the design to a variety of engineering analyses and can identify potential problems (such as an excessive load or deflection). The speed and accuracy of such analyses far surpass what is available from traditional methods. The CAD user inputs data by keyboard and/or mouse to produce illustrations on the computer screen that can be reproduced as paper copies with a plotter (see Fig. 12.3) or printer.

Draft efficiency is significantly improved. When something is drawn once, it never has to be drawn again. It can be retrieved from a library, and can be duplicated, stretched, sized, and changed in many ways without having to be redrawn. Cut and paste techniques are used as labor-saving aids.

MACHINE COMPONENTS AND DESIGN

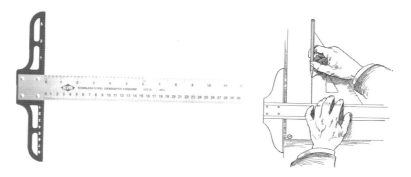

Figure 12.1 T-squares

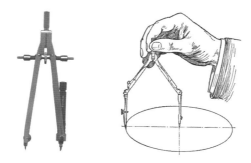

Figure 12.2 Drawing compasses

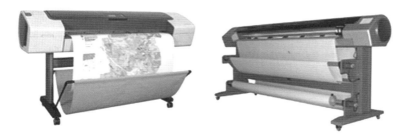

Figure 12.3 Plotters

Engineers generally agree that the computer does not change the nature of the design process but is a significant tool that improves efficiency and productivity. The designer and the CAD system may be described as a design team: the designer provides knowledge, creativity, and control; the computer generates accurate, easily modifiable graphics, performs complex design analysis at great speed, and stores and recalls design information. Occasionally, the computer may augment or replace many of the engineer's other tools, but it cannot replace the design process, which is controlled by the designer.

Depending on the nature of the problem and the sophistication of the computer system, computers offer the designer or drafter some or all of the following advantages.

1. Easier creation and correction of drawings. Engineering drawings may be created more quickly than by hand and making changes and modifications is more efficient than correcting drawings made by hand.

2. Better visualization of drawings. Many systems allow different views of the same object to be displayed and 3D pictorials (see Fig. 12.4) to be rotated on the computer screen.

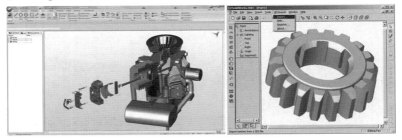

Figure 12.4   3D pictorials

3. Database of drawing aids. Creation and maintenance of design databases (libraries of designs) permits storing designs and symbols for easy recall and application to the solution of new problems.

4. Quick and convenient design analysis. Because the computer offers ease of analysis, the designer can evaluate alternative designs, thereby considering more possibilities while speeding up the process at the same time.

5. Simulation and testing of designs. Some computer systems make possible the simulation of a product's operation, testing the design under a variety of conditions and stresses. Computer testing may improve on or replace construction of models and prototypes.

6. Increased accuracy. The computer is capable of producing drawings with more accuracy than is possible by hand. Many CAD systems are even capable of detecting errors and informing the user of them.

7. Improved filing. Drawings can be more conveniently filed, retrieved, and transmitted on disks and tapes.

Computer graphics has an almost limitless number of applications in engineering and other technical fields. Most graphical solutions that are possible with a pencil can be done on a computer and usually more productively. Applications vary from 3D modeling and finite element analysis to 2D drawings and mathematical calculations.

Once the domain of large computer systems advanced applications can now be done on microcomputers. An important extension of CAD is its application to manufacturing. Computer-aided design/computer-aided manufacturing (CAD/CAM) systems may be used to design a part or product, devise the essential production steps, and electronically communicate this information to and control the operation of manufacturing equipment, including robots. These systems offer many advantages over traditional design and manufacturing systems, including less design effort, more efficient material use, reduced lead time, greater accuracy, and improved inventory control.

Words and Expressions

literate ['litərit] a. 有文化的,有阅读和写作能力的
engineering drawing 工程图样,工程制图
graphic image 图形,图像
plotter ['plɔtə] n. 绘图机,绘图仪
retrieval [ri'tri:vəl] n. 检索,从内存或其它存储设备中获取信息的过程
offshoot ['ɔ:ʃu:t] n. 分支,支流
keyboard and mouse 键盘和鼠标
monitor ['mɔnitə] screen 显视器屏幕
paper copy 打印输出
recall [ri'kɔ:l] v. 检索,再调用
augment [ɔ:g'ment] v. 增加,增大; n. 增加
3D pictorial 三维图像
library of design 设计方案库
alternative design 替代设计(方案),可供选择的设计(方案)
simulation [,simju'leiʃən] n. 仿真,模拟
prototype ['prəutətaip] n. 原型,样机
filing ['failiŋ] n. 整理成档案,文件归档,存档
3D modeling 三维建模,创建三维模型,三维造型
finite element analysis 有限元分析
design effort 设计工作
lead time 产品的研制周期(从设计到实际投产),从订货到交货的时间间隔
inventory control 库存管理

13. Responsibility, Liability, and Litigation

Engineers are responsible at the very least to adhere to a code of professional ethics. The following statements, prepared by the Ethics Committee of the older ECPD (Engineer's Council for Professional Development), the predecessor of ABET (Accreditation Board for Engineering and Technology), continue to provide a simple, if not idealistic, framework:

Faith of the engineer

"I am an engineer. In my profession I take deep pride, but without vain-glory; to it I owe solemn obligations that I am eager to fulfill.

As an Engineer, I will participate in none but honest enterprise. To him that has engaged my services, as employer or client, I will give the utmost of performance and fidelity.

When needed, my skill and knowledge shall be given without reservation for the public good. From special capacity springs the obligation to use it well in the service of humanity; and I accept the challenge that this implies.

Jealous of the high repute of my calling, I will strive to protect the interests and the good name of any engineer that I know to be deserving; but I will not shrink, should duty dictate, from disclosing the truth regarding anyone that, by unscrupulous act, has shown himself unworthy of the profession.

Since the Age of Stone, human progress has been conditioned by the genius of my professional forbears. By them have been rendered usable to mankind Nature's vast resources of material and energy. By them have been vitalized and turned to practical account the principles of science and revelations of technology. Except for this heritage of accumulated experience, my efforts would be feeble. I dedicate myself to the dissemination of engineering knowledge, and, especially to the instruction of younger members of my profession in all its arts and traditions.

To my fellows I pledge, in the same full measure I ask of them, integrity and fair dealing, tolerance and respect, and devotion to the standards and the dignity of our profession; with the consciousness, always, that our special expertness carries with it the obligation to serve humanity with complete sincerity. "

Liability concepts

Liability in its simplest terms is a condition of being responsible for an actual or possible loss, penalty, or burden. Thorpe and Middendorf, in their treatment of liability, describe contractual and tort concepts. In the former, a contract, warranty, or guarantee can represent an agreement that is expressed (usually in writing). Failure to comply with this agreement or a violation of this agreement constitutes a liability. There is also an implied warranty that carries the expectation that any product or component offered for sale is "reasonably" safe. In the latter concept, a tort represents a wrongful or "irresponsible" act that results in injury (even to a successful business, for example) and from which some civil legal action (litigation) may result. To establish liability in the tort concept, it is therefore necessary to establish negligence (when the standard is based on what a "reasonable" engineer, designer, or manufacturer would have done), which is referred to as absolute liability, or to demonstrate strict liability. In strict liability, it is not necessary to demonstrate negligence on the part of the manufacturer, engineer, or designer but simply to demonstrate that a product or component was defective, unreasonably dangerous, and the proximate cause of injury.

As a matter of professional responsibility and as a means to reduce the risk of liability, it is incumbent upon every engineer to

(1) Conform to the highest standards of his or her field; to keep the faith.

(2) Follow all safety requirements, design codes, standards, and practices in a responsible fashion. Don't take reckless shortcuts, and don't be pressured by anyone into abandoning your professional principles. Have the guts to express a concern for safety even if you may be wrong.

(3) If compromises must be made in product design or development, clearly state the limitations of use. Liability is often demonstrated simply as a failure to reasonably inform a user of a product's limitations.

(4) Consider potential design flaws or materials failure in light of a careless or reckless user.

Litigation

Litigation is the act of carrying on a suit or claim in a court of law; a judicial contest involving any controversy that must be decided upon evidence. Litigation involves the process of deciding liability. In the case of product failure or material or component failure, failure analysis provides evidence. Engineers actually provide such evidence; attorneys present the evidence and argue for or against liability on the basis of how they interpret the evidence.

Words and Expressions

at the very least [用于加强语气] = at (the) least 至少,起码,无论如何

adhere to 坚持,遵守

professional ethics 职业道德
council ['kaunsl] n. 委员会,参议会,政务会
predecessor ['pri:disesə] n. 前辈,(被替代的)原有(事)物
framework ['freimwə:k] n. 构架,框架,结构
vain-glory [vein'glɔ:ri] n. 自负,虚荣心
solemn ['sɔləm] a. 庄严的,严肃的
engage [in'geidʒ] v. 使从事于,参加
utmost ['ʌtməust] n. 极限,最大可能的; a. 极度的
fidelity [fai'deliti] n. 忠诚,忠实
jealous ['dʒeləs] a. 妒忌的,注意的,唯恐失掉的
calling ['kɔ:liŋ] n. 职业,行业,名称
dictate ['dikteit] v. 指示,命令,规定,要求
disclose [dis'kləuz] v. 揭露,揭发,揭开,泄露
unscrupulous [ʌn'skru:pjuləs] a. 不审慎的,不讲道德的
Age of Stone 石器时代
forbear [fɔ:'bɛə] n. 祖先,前辈
vitalize ['vaitəlaiz] v. 使有生命力,使增添活力,激发
turn to...account 利用
revelation [revi'leiʃən] n. 新发现,展现
heritage ['heritidʒ] n. 遗产,继承物,传统
dissemination [disemi'neiʃən] n. 散布,传播(思想,理论等)
integrity [in'tegriti] n. 完整性,完善,正直,诚实
fair dealing 公平交易
sincerity [sin'seriti] n. 真实,真诚,诚挚
in their treatment of liability 在他们关于赔偿责任的论述中
contractual [kən'træktjuəl] a. 契约的,合同的
guarantee [gærən'ti:] n.; v. 保证(书,人),担保,承诺
irresponsible [iris'pɔnsəbl] a. 不负责任的,不可靠的
civil legal action 民事诉讼
absolute liability 绝对责任,绝对赔偿责任

strict liability 严格赔偿责任
proximate ['prɔksimət] a. 最近的,直接的
the proximate cause 直接原因
conform [kən'fɔ:m] v. 使一致,使遵守,符合
conform to 符合,遵照
reckless ['reklis] a. 粗心大意的鲁莽的轻率的,不计后果的
shortcut ['ʃɔ:tkʌt] n. 捷径
abandon [ə'bændən] v. 放弃,废弃,舍弃
have the guts to do sth 有做某事的勇气
carry on 开发,从事,维持,坚持
suit [su:t] n. 控告,起诉,诉讼
judicial [dʒu:'diʃəl] a. 司法的,法院的,公正的,明断的
contest [kən'test] v. ['kɔntest] n. 争论,辩论,比赛
controversy ['kɔntrəvə:si] n. 争论,论战
in the case of 在…的情况下,就…来说
for or against 赞成或反对

2 MACHINE TOOLS AND MACHINING

14. Engine Lathes

The engine lathe, one of the oldest metal removal machines, has a number of useful and highly desirable attributes. Today these lathes are used primarily in small shops where smaller quantities rather than large production runs are encountered.

The essential components of an engine lathe are the bed, headstock, tailstock, carriage, leadscrew and feed rod, as shown in Fig. 14.1.

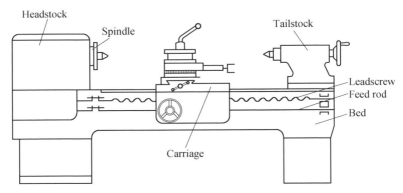

Figure 14.1 Schematic layout of an engine lathe

The bed is the backbone of an engine lathe. It is usually made of well-normalized or aged gray or nodular cast iron and provides a heavy, rigid frame on which all the other basic components are mounted. Two sets of parallel, longitudinal ways, inner and outer, are contained on the bed, usually on the upper side. Some makers use an inverted V-shape for all four ways, whereas others utilize one inverted V and one flat way in one or both sets (see Fig. 14. 2). They are precision-machined to assure accuracy of alignment. On most modern lathes the ways are surface-hardened to resist wear and abrasion, but precaution should be taken in operating a lathe to assure that the ways are not damaged. Any inaccuracy in them usually means that the accuracy of the entire lathe is destroyed.

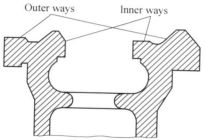

Figure 14. 2　Shape of lathe ways

The headstock is mounted in a fixed position on the inner ways, usually at the left end of the bed. It provides a powered means of rotating the work at various speeds. Essentially, it consists of a hollow spindle, mounted in accurate bearings, and a set of transmission gears—similar to a truck transmission—through which the spindle can be rotated at a number of speeds. Most lathes provide from 8 to 18 speeds, usually in a geometric ratio. An increasing trend is to provide a continuously variable speed range through electrical or mechanical drives.

Because the accuracy of a lathe is greatly dependent on the spindle, it is of heavy construction and mounted in heavy bearings, usually preloaded tapered roller or ball types (see also Figs. 3.2b and 3.2c). The spindle has a hole extending through its length, through which long bar stock can be fed. The size of this hole is an important dimension of a lathe because it determines the maximum size of bar stock that can be machined when the material must be fed through spindle.

The chuck is connected to the spindle, where clamps and rotates the work. The three-jaw "self-centering" chuck (Fig. 14.3a) moves all of its jaws simultaneously to clamp or unclamp the work, and it is used for work with a round or hexagonal cross-section. The four "independent jaw" chuck (Fig. 14.3b) can clamp on the work by moving each jaw independent of the others. This chuck exerts a stronger hold on the work and it has the ability to center non-round shapes (squares, rectangles) exactly.

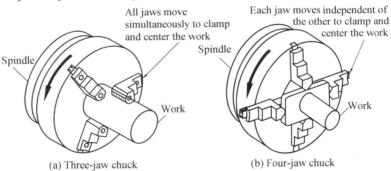

(a) Three-jaw chuck　　　　(b) Four-jaw chuck

Figure 14.3　Chucks

The tailstock is towards the right-most end on the bed, which provides a tailstock quill for the purpose of locating the long workpieces by the use of centers. The tailstock is movable on the inner ways of the bed to accommodate the different lengths of workpieces. It also serves the purpose of holding tools such as center drill, twist drill, reamer, etc. for making and finishing holes in the workpieces which are located in line with the axis of rotation.

The three-jaw and four-jaw chucks are normally suitable for short workpieces. Workpieces that are relative long with respect to their diameters are machined between centers. Before a workpiece can be mounted between lathe centers, a 60° center hole must be drilled in each end (see Fig. 14.4). Through these center holes the centers mounted in the spindle and the tailstock would locate the axis of the workpiece. However, these centers would not be able to transmit the motion to the workpiece from the spindle. For this purpose, generally a driver plate and a lathe dog would be used. In Fig. 14.4, the center located in the spindle is a dead center while that in the tailstock is a live center (see Fig. 14.5). The shank of the center is generally finished with a Morse taper which fits into the tapered hole of the spindle or tailstock.

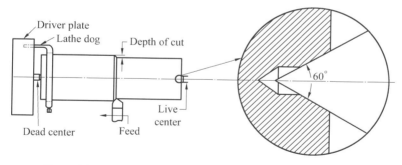

Figure 14.4 Workpiece being turned between centers in a lathe

The size of a lathe is designated by two dimensions. The first is known as the swing. It is approximately twice the distance between the line connecting the lathe centers and the nearest point on the ways. The second size dimension is the maximum distance between centers. The swing thus indicates the maximum workpiece diameter that can be turned in the lathe, while the distance between centers indicates the maximum length of workpiece that can be mounted between centers.

(a) Dead center (b) Live center

Figure 14.5　Centers

Words and Expressions

large production runs 大批量生产
headstock[ˈhedstɔk] n. 主轴箱,动力箱
tailstock[ˈteilstɔk] n. 尾座,尾架
carriage[ˈkæridʒ] n. 溜板(在床身上使刀具作纵向移动的部件,一般由刀架、床鞍、溜板箱等组成)
leadscrew [ˈliːdskruː] n. 丝杠
feed rod 光杠
schematic layout 示意图,简图
way 导轨
surface-hardened 表面淬火
abrasion[əˈbreiʒən] n. 擦伤,刮去,磨损
geometric ratio 等比,等比级数
bar stock 棒料
preload[ˈpriːˈləud] n. ;v. 预加荷载
tapered[ˈteipəd] a. 锥形的,斜的
longitudinally[lɔndʒiˈtjuːdinli] ad. 长度地,纵向地,轴向地
thereon[ðɛəˈɔn] ad. 在其中,在其上,关于那,紧接着
quill [kwil] n. 活动套筒,衬套,钻轴,空心轴
by the use of 通过利用

center ['sentə] n. 中心, 顶尖; v. 定心, 对中
center hole 中心孔, 顶尖孔
center drill 中心钻
axis ['æksis] n. 轴, 轴线
driver plate 拨盘, 有时也写为 drive plate
lathe dog 卡箍, 鸡心夹头, 有时也写为 dog
dead center 普通顶尖, 死顶尖, 固定顶尖
live center 活顶尖, 回转顶尖
Morse taper 莫氏锥度
swing [swiŋ] n. 最大回转直径
center distance 中心距, 顶尖距

15. Milling Operations

After lathes, milling machines are the most widely used for manufacturing applications. In milling, the workpiece is fed into a rotating milling cutter, which is a multi-edge tool as shown in Fig. 15.1. Metal removal is achieved through combining the rotary motion of the milling cutter and linear motions of the workpiece simultaneously.

Each of the cutting edges of a milling cutter acts as an individual single-point cutter when it engages with the workpiece metal. Therefore, each of those cutting edges has appropriate rake and relief angles. Since only a few of the cutting edges are engaged with the workpiece at a time, heavy cuts can be taken without adversely affecting the tool life. In fact, the permissible cutting speeds and feeds for milling are three to four times higher than those for turning or drilling. Moreover, the quality of surfaces machined by milling is generally superior to the quality of surfaces machined by turning,

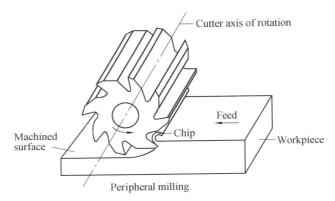

Figure 15.1 Schematic diagram of a milling operation

shaping, or drilling.

As far as the directions of cutter rotation and workpiece feed are concerned, milling is performed by either of the following two methods.

(1) **Up milling** (**conventional milling**). In up milling the cutting tool rotates in the opposite direction to the table movement, the chip starts as zero thickness and gradually increases to the maximum size as shown in Fig. 15.2a. This tends to lift the workpiece from the table. There is a possibility that the cutting tool will rub the workpiece before starting the removal. However, the machining process involves no impact loading, thus ensuring smoother operation of the machine tool.

The initial rubbing of the cutting edge during the start of the cut in up milling tends to dull the cutting edge and consequently have a lower tool life. Also since the cutter tends to cut and slide alternatively, the quality of machined surface obtained by this method is not very high. Nevertheless, up milling is commonly used in industry, especially for rough cuts.

(2) **Down milling** (**climb milling**). In down milling the cutting

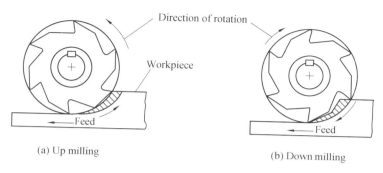

(a) Up milling (b) Down milling

Figure 15.2 Milling methods

tool rotates in the same direction as that of the table movement, and the chip starts as maximum thickness and goes to zero thickness gradually as shown in Fig. 15.2 b.

The advantages of this method include higher quality of the machined surface and easier clamping of workpieces, since cutting forces act downward. Down milling also allows greater feeds per tooth and longer cutting life between regrinds than up milling. But, it cannot be used for machining castings or hot rolled steel, since the hard outer scale will damage the cutter.

There are a large variety of milling cutters to suit specific requirements. The cutters most generally used, shown in Fig. 15.3, are classified according to their general shape or the type of work they will do.

Plain milling cutters. They are basically cylindrical with the cutting teeth on the periphery, and the teeth may be straight or helical, as shown in Fig. 15.3a. The helical teeth generally are preferred over straight teeth because the tooth is partially engaged with the workpiece as it rotates. Consequently, the cutting force variation will be smaller, resulting in a smoother operation. The cutters are generally used for machining flat surfaces.

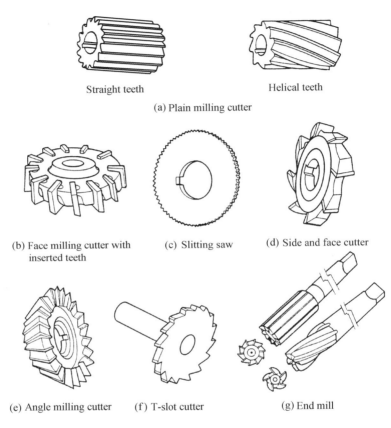

Figure 15.3 Types of milling cutters

Face milling cutters. They have cutting edges on the face and periphery. The cutting teeth, such as carbide inserts, are mounted on the cutter body as shown in Fig. 15.3b.

Most larger-sized milling cutters are of inserted-tooth type. The cutter body is made of ordinary steel, with the teeth made of high speed steel, cemented carbide, or ceramics, fastened to the cutter body by various methods. Most commonly, the teeth are indexable carbide or ceramic inserts.

Slitting saws. They are very similar to a saw blade in appearance as well as function (Fig. 15.3c). The thickness of these cutters is generally very small. The cutters are employed for cutting off operations and deep slots.

Side and face cutters. They have cutting edges not only on the periphery like the plain milling cutters, but also on both the sides (Fig. 15.3d). As was the case with the plain milling cutter, the cutting teeth can be straight or helical.

Angle milling cutters. They are used in cutting dovetail grooves and the like. Figure 15.3e indicates a milling cutter of this type.

T-slot cutters. T-slot cutters (Fig. 15.3f) are used for milling T-slots such as those in the milling machine table.

End mills. There are a large variety of end mills. One of distinctions is based on the method of holding, i.e., the end mill shank can be straight or tapered. The straight shank is used on end mills of small size. The tapered shank is used for large cutter sizes. The cutter usually rotates on an axis perpendicular to the workpiece surface.

Figure 15.3g shows two kinds of end mills. The cutter can remove material on both its end and its cylindrical cutting edges. Vertical milling machines and machining centers can be used for end milling workpieces of various sizes and shapes. The machines can be programmed such that the cutter can follow a complex set of paths that optimize the whole machining operation for productivity and minimum cost.

Words and Expressions

milling cutter 铣刀
multi-edge tool 多刃刀具(有多个主切削刃参加切削的刀具),

也可以写为 multi-point tool
machined surface 已加工表面
peripheral milling 圆周铣削,周铣
schematic [ski'mætik] a. 示意性的,图解的,图表的
schematic diagram 原理图,示意图
single-point cutter 单刃刀具(切削时只用一个主切削刃的刀具),也可以写为 single-point tool
cutting edge 切削刃,刀刃
rake angle 前角
relief angle 后角
chip [tʃip] n. 切屑
heavy cut 重切削,强力切削
up milling 逆铣
conventional milling 逆铣
down milling 顺铣
climb milling 顺铣
table 工作台
table feed 工作台进给
cutting life 刀具寿命
regrind [ri'graind] v. 重新研磨,再次磨削
casting ['kɑːstiŋ] n. 铸件,铸造
hot rolled steel 热轧钢
outer scale 外层氧化皮
helical teeth 螺旋齿
plain milling cutter 圆柱铣刀
face milling cutter 面铣刀,端铣刀
face milling cutter with inserted teeth 镶齿面铣刀,镶齿端铣刀
slitting saw 锯片铣刀
cut off 切断
side and face cutter 三面刃铣刀

angle milling cutter 角度铣刀
T-slot cutter　T形槽铣刀
end mill 立铣刀
cutter body 刀体
cemented carbide 硬质合金
fasten[′fɑːsn] v. 固定,使坚固或稳固
saw blade 锯条
carbide insert 硬质合金刀片
indexable ceramic insert 陶瓷可转位刀片
as was the case with 正如,就像
dovetail groove 燕尾槽
and the like 等等
tapered shank 锥柄
machining center 加工中心

16. Drilling Operations

Drilling involves producing through or blind holes in a workpiece by forcing a tool, which rotates around its axis, against the workpiece. Consequently, the range of cutting from that axis of rotation is equal to the radius of the required hole.

Cutting Tools for Drilling Operations

In drilling operations, a cylindrical cutting tool, called a drill, is employed. The drill can have either one or more cutting edges and corresponding flutes, which can be straight or helical. The function of the flutes is to provide outlet passages for the chips generated during the drilling operation and also to allow lubricants and coolants to reach the cutting edges and the surface being machined. Following is a survey of the commonly used drills.

Twist Drills The twist drill is the most common type of drill. It has two cutting edges and two helical flutes that continue over the length of the drill body. The shank of a drill can be either straight (Fig. 16.1a) or tapered (Fig. 16.1b). In the latter case, the shank is fitted into the tapered socket of the spindle and has a tang, which goes into a slot in the spindle socket, thus acting as a solid means for transmitting rotation.

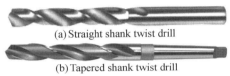

(a) Straight shank twist drill

(b) Tapered shank twist drill

Figure 16.1 Twist drills

Core Drills A core drill consists of the chamfer, body, neck, and shank, as shown in Fig. 16.2. This type of drill may have either three or four flutes and an equal number of margins, which ensure superior guidance, thus resulting in high machining accuracy. A core drill has flat end. The chamfer can have three or four cutting edges. Core drills are employed for enlarging previously made holes and not for originating holes. This type of drill is characterized by greater productivity, high machining accuracy, and superior quality of the drilled surfaces.

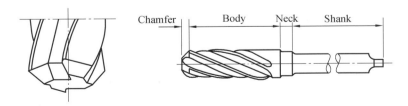

Figure 16.2 Core drill

Carbide Tipped Drills Most of the drills are made of high speed steel. However, for machining hard materials as well as for large volume production, carbide tipped drills are available. As shown in Fig. 16.3, the carbide tips of suitable geometry are clamped to the end of the tool to act as the cutting edges.

Figure 16.3 Drill with carbide inserts

Other Types of Drilling Operations

In addition to conventional drilling, there are other operations that are involved in the production of holes in the industrial practice. Following is a brief description of each of these operations.

Boring Boring involves enlarging a hole that has already been drilled. It is similar to internal turning and can, therefore, be performed on a lathe. There are also some specialized machine tools for carrying out boring operations. Those include the vertical boring machine, the jig boring machine, and the horizontal boring machine.

Reaming Reaming is an operation used to make an existing hole dimensionally more accurate than can be obtained by drilling alone. As a result of a reaming operation, a hole has a very smooth surface. The cutting tool used in this operation is known as the reamer (Fig. 16.4).

Tapping Tapping is the process of cutting internal threads. The tool is called a tap (Fig. 16.5).

Figure 16.4 Reamers   Figure 16.5 Taps

Classification of Drilling Machines

Drilling operations can be carried out by using either electric hand drills or drilling machines. The latter differ in shape and size. Nevertheless, the tool always rotates around its axis while the workpiece is kept firmly fixed. This is contrary to the drilling operation on a lathe, where the workpiece is held in and rotates with the chuck. Following is a survey of the commonly used types of drilling machines.

Bench-type Drilling Machines Bench-type drilling machines are general-purpose, small machine tools that are usually placed on benches. This type of drilling machine includes an electric motor as the source of motion, which is transmitted via pulleys and belts to the spindle, where the tool is mounted (see Fig. 16.6). The feed is manually generated by lowering a feed lever, which is designed to lower (or raise) the spindle. The workpiece is mounted on the machine table, although a special vise is sometimes used to hold the workpiece.

Upright Drilling Machines Depending upon the size, upright drilling machine tools can be used for light, medium, and even relatively heavy workpieces. It is basically similar to bench-type machines, the main difference being a longer cylindrical column fixed to the base. Along that column is an additional, sliding table for fixing the workpiece which can be locked in position at any desired height.

MACHINE TOOLS AND MACHINING 83

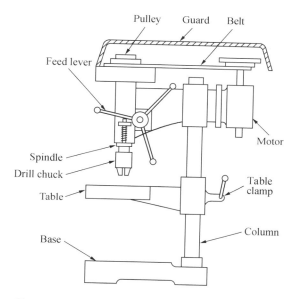

Figure 16.6 Sketch of a bench-type drilling machine

The power required for this type is more than that for the bench-type drilling machines.

Radial Drills A radial drill is particularly suitable for drilling holes in large and heavy workpieces that are inconvenient to mount on the table of an upright drilling machine. A radial drilling machine has a main column, which is fixed to the base. The cantilever guide arm, which carries the drilling head, can be raised or lowered along the column and clamped at any desired position (see Fig. 16.7). The drilling head slides along the arm and provides rotary motion and axial feed motion. Again, the cantilever guide arm can be swung, thus enabling the tool to be moved in all directions according to a cylindrical coordinate system.

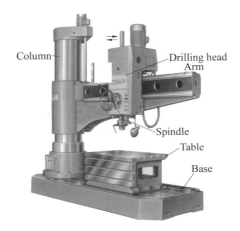

Figure 16.7 Radial drill

Words and Expressions

drilling operation 钻孔加工,钻削加工
through or blind holes 通孔或盲孔
symmetrical[si'metrikəl] a. 对称的,平衡的
flute[fluːt] n. 凹槽,(刀具的)排屑槽,容屑槽
outlet['autlet] n. 排出口,流出口,排泄口,排水孔
coolant['kuːlənt] n. 冷却液,切削液,乳化液
twist drill 麻花钻头
shank[ʃæŋk] n. 刀柄,尾部,后部,杆
tapered socket of the spindle 主轴锥孔
core drill 扩孔钻
chamfer['tʃæmfə] n. 倒角,斜面,(扩孔钻的)切削部分
margin['mɑːdʒin] n. 刃带
guidance['ɡaidəns] n. 引导,制导,向导,引导装置
enlarge[in'lɑːdʒ] v. 扩大,扩充,增大
originate[ə'ridʒineit] v. 起源,发源,引起,产生

machining accuracy 机械加工精度
carbide['kɑ:baid] n. 碳化物,硬质合金
tip[tip] n. 尖,梢,端,刀片
carbide tipped drill 镶硬质合金钻头,硬质合金钻头
large volume production 大批量生产
indexable insert 可转位刀片
tapping['tæpiŋ] n. 攻螺纹
tap[tæp] n. 丝锥
vertical boring machine 立式镗床
jig boring machine 坐标镗床
horizontal boring machine 卧式镗床
electric hand drill 手电钻
bench-type drilling machine 台式钻床
general purpose 通用的,多种用途的
upright['ʌp'rait] a. ;ad. 笔直的,竖立的,n. 支柱
upright drilling machine 立式钻床
preset['pri:'set] v. ;a. 预调,预先设置
vise[vais] n. 虎钳,台钳;v. 钳住,夹紧
column['kɔləm] n. 柱,柱状物,架,墩,(钻床等)床身
medium-duty 中型的,中等的,中批生产
radial drill 摇臂钻床
drilling head 钻床主轴箱,钻削动力头
cantilever['kæntili:və] n. 悬臂;v. 使……伸出悬臂梁
cylindrical coordinate system 柱面坐标系统

17. Grinding Machines

Grinding is a chip-making metalcutting process, every bit as much as turning or milling, except that it's performed on a "micro" level where the chips are too small to be easily seen and identified. Another basic difference between grinding and its large-scale cousins, is that the grinding "tools", which are abrasive particles, are not uniformly shaped and oriented in the carrier, be it a wheel, stone, or some other device.

In metalcutting terms, grinding is a hybrid between turning and milling. Like milling, the grinding tool is in motion, and like turning, the workpiece is also nearly always moving, although not necessarily rotating. That distinction between rotating workpieces, and those moving in a linear or oscillating fashion under the grinding tool, but otherwise fixed, defines the most basic division in grinding applications. The former is generically described as "cylindrical" grinding, while the latter is "surface" grinding, even though both types can produce precision surfaces on the workpiece which aren't necessarily round or flat.

The fact that both the tool and workpiece are normally moving makes analysis and control of grinding operations considerably more complex than is the case with turning or milling. This is one of the main reasons why many grinder manufacturers still tend to design and build their own CNC controls.

Many cylindrical grinders are laid out on a pattern similar to a lathe. Instead of a toolholder, they have a mechanism to feed the grinding wheel, but all the other elements, headstock, tailstock, ways, and bed are normally quite recognizable to anyone familiar with turning machines. The major difference, however, is that the entire workholding system usually is mounted on a set of ways so it can be traversed past the wheel during the grinding operation.

Typically, in an OD cylindrical grinder the wheelhead is fed radially into the workpiece which may either remain stationary or be reciprocated past it. When the workpiece reciprocates, infeed motions are typically made only at the end of each stroke.

The third major type of cylindrical grinder is the "centerless" design, which has no direct analog in the world of turning. In a centerless grinder, the workpiece is not held by either centers or a chuck, but is supported on a "blade" between the grinding wheel and a "regulating" wheel which controls its rotation.

The "regulating" wheel controls the rotational speed of the workpiece, serving as both a driving mechanism, and a brake if the workpiece picks up speed from the grinding wheel. The "blade" keeps the work in position between the two wheels(see Fig. 17.1).

Figure 17.1　Centerless grinding

Surface grinders also come in several basic designs. Probably the most common has the wheel and its associated machinery mounted above a reciprocating table or other linear motion device that holds the workpiece. The wheel is fed across the axis of reciprocation, while the workpiece moves back and forth under it.

Conceptually, this type of grinder is similar to a planer in which the toolholder has been replaced by the grinding wheel. The grinding is normally done with the periphery of the wheel on this type of machine.

Grinding is a process with such a broad application range that many variations on these basic design themes have been developed over the years, and no attempt will be made to cover them all here in detail. Jig grinders, ultra-precision bearing ball grinders, and a whole spectrum of other specialized grinders could each easily be the topic of an article longer than this one. The principles, however, are essentially universal and should be applicable with slight modification to virtually any grinding application you may have.

One type of specialty grinder that does merit brief mention is the tool and cutter grinder which, as its names implies, is designed for sharpening and re-sharpening cutting tools. These machines are essentially hybrids incorporating features of both cylindrical and surface grinders which are necessary for the specialized tasks they are required to perform.

Regardless of the type of grinder, the most important attributes it must have are rigidity and stability. This is particularly true in creep feed applications, a special form of grinding which is performed in a single pass, with a very high depth of cut, and low workspeed. Properly applied, creep feed grinding can cut overall machining time by up to 50% with no loss of dimensional or geometric precision or surface finish quality.

To achieve these results, however, the grinder must be designed specifically for creep feed applications, since the technique is particularly sensitive to the static and dynamic stability of the machine and requires up to three times the spindle power of a comparable conventional grinding process. Creep feed also requires special dressing capabilities, careful attention to wheel "hardness", good coolant control, and a great deal of experience-based process knowledge.

Creep feed, like all other metalworking technologies, offers significant advantages for specific applications when properly applied. It is not a panacea, and it is not the answer to every grinding problem.

In view of its rather specialized nature, you are well advised to approach the process with a healthy caution to make certain it's the right answer to your needs before investing your money.

As was the case with turning machines there really aren't any secrets in the structural design of grinders. You are somewhat more likely to see stability-enhancing/vibration-damping features like polymer or concrete-filled bases in grinders than in turning machines or machining centers, but, by and large, the structural choices you will be presented with will be familiar and relatively easy to sort out.

Words and Expressions

machining [məˈʃiːniŋ] n. 机械加工, 切削加工
grinding [ˈgraindiŋ] n. 磨削; a. 磨削的
chip [tʃip] n. 切屑, 金属屑, 芯片
every bit 每一点, 完全
as much as 差不多, …那样多
turning [ˈtəːniŋ] n. 旋转, 车削
milling [ˈmiliŋ] n. 铣削
micro [ˈmaikrəu] n. ; a. 微型(的), 微细(的), 微观(的)
large scale 大比例的, 大规模的
orient [ˈɔːriənt] v. 确定方向
carrier [ˈkæriə] n. 载体, 承载部件
purpose-built 为特定目的而建造的, 特别的, 专用的
metalworking [ˈmetlwəːkiŋ] n. 金属加工, 金工
hybrid [ˈhaibrid] n. 混合物; a. 混合式的
linear [ˈliniə] a. 线的, 直线的, 线性的
oscillate [ˈɔsileit] v. (来回)摆动, 振荡, 摇摆
cylindrical grinding 圆柱面磨削, 外圆磨削
surface grinding 平面磨削

CNC = computer numerical control 计算机数字控制
toolholder [ˈtuːlhəuldə] n. 刀夹(用来安装和紧固切刀的工具)
mechanism [ˈmekənizəm] n. 机构,机械装置
feed [fiːd] (fed,fed) v. ;n. 进给,进给量
headstock [ˈhedstɔk] n. 头架,主轴箱
tailstock [ˈteilstɔk] n. 尾架,尾座
ways and bed 导轨和床身
a set of 一组,一套
traverse [ˈtrævəːs] v. 横向移动
OD = outside diameter 外径
OD cylindrical grinder 外圆磨床
wheelhead 砂轮头
radially [ˈreidiəli] ad. 径向地
reciprocate [riˈsiprəkeit] v. (使)往复运动
infeed 横向进磨,切入磨法,横切(进给)
stroke [strəuk] n. 行程
centerless grinder 无心磨床
blade (无心磨床上的)托板
regulating wheel 导轮
brake [breik] n. 制动器,制动装置;v. 制动,刹车
pick up 加快,加速
planer [ˈpleinə] n. (龙门)刨床
periphery [pəˈrifəri] n. 周边,圆周,圆柱(体)表面
jig grinder 坐标磨床
tool and cutter grinder 工具磨床
rigidity [riˈdʒiditi] n. 刚性,刚度,坚固性
creep feed grinding 缓进给磨削
workspeed = workpiece speed 工件速度
dressing [ˈdresiŋ] n. (砂轮)修整(产生锋锐的磨削刃)
panacea [pænəˈsiə] n. 灵丹妙药,解决一切问题的方法

healthy 康健的,大量的
machining center 加工中心
by and large 大体上,基本上
sort out 把…分类,选出,拣出

18. Machining

Machining can be defined as the process of removal of the unwanted material (machining allowance) from the workpiece in the form of chips, so as to obtain a finished product of the desired size, shape, and surface quality. Machining includes turning, milling, grinding, drilling, etc.

Turning

Engine Lathes The engine lathe, one of the oldest metal removal machines, has a number of useful and highly desirable attributes. Today these lathes are used primarily in small shops where smaller quantities rather than large production runs are encountered.

Tolerances for the engine lathe depend primarily on the skill of the operator. The design engineer must be careful in using tolerances of an experimental part that has been produced on the engine lathe by a skilled operator. In redesigning an experimental part for production, economical tolerances should be used.

The engine lathe has been replaced in today's production shops by a wide variety of automatic lathes such as turret lathes, automatic screw machines, and automatic tracer lathes.

Turret Lathes Production machining equipment must be evaluated now, more than ever before, in terms of ability to repeat accurately and rapidly. Applying this criterion for establishing the production qualification of a specific method, the turret lathe merits a high rating.

In designing for low quantities such as 100 or 200 parts, it is most economical to use the turret lathe. In achieving the optimum tolerances possible on the turret lathe, the designer should strive for a minimum of operations.

Automatic Tracer Lathes Since surface roughness depends greatly upon material turned, tooling, and feeds and speeds employed, minimum tolerances that can be held on automatic tracer lathes are not necessarily the most economical tolerances.

In some cases, tolerances of ±0.05 mm are held in continuous production using but one cut. Groove width can be held to ±0.125 mm on some parts. On high production runs where maximum output is desirable, a minimum tolerance of ±0.125 mm is economical on both diameter and length of turn.

Milling

With the exceptions of turning and drilling, milling is undoubtedly the most widely used method of removing metal. Well suited and readily adapted to the economical production of any quantity of parts, the almost unlimited versatility of the milling process merits the attention and consideration of designers seriously concerned with the manufacture of their product.

As in any other process, parts that have to be milled should be designed with economical tolerances that can be achieved in production milling. If the part is designed with tolerances finer than necessary, additional operations will have to be added to achieve these tolerances-and this will increase the cost of the part.

Grinding

Grinding is one of the most widely used methods of finishing parts to extremely close tolerances and fine surface finishes. Currently, there are grinders for almost every type of grinding operation. Particular design features of a part dictate to a large degree the type of grinding machine required. Where processing costs are excessive, parts redesigned to utilize a less expensive, higher output grinding method may be well worthwhile. For example, wherever possible the production economy of centerless grinding should be taken advantage of by proper design consideration.

Although grinding is usually considered a finishing operation, it is often employed as a complete machining process on work which can be ground down from rough condition without being turned or otherwise machined. Thus many types of forgings and other parts are finished completely with the grinding wheel at appreciable savings of time and expense.

Classes of grinding machines include the following: cylindrical grinders, centerless grinders, internal grinders, surface grinders, and tool and cutter grinders.

The cylindrical and centerless grinders are for straight cylindrical or taper work; thus splines, shafts, and similar parts are ground on cylindrical machines either of the common-center type or the centerless machine.

Thread grinders are used for grinding precision threads for thread gages, and threads on precision parts where the concentricity between the diameter of the shaft and the pitch diameter of the thread must be held to close tolerances.

The internal grinders are used for grinding of precision holes, cylinder bores, and similar operations where bores of all kinds are to be finished.

The surface grinders are for finishing all kinds of flat work, or work with plain surfaces which may be operated upon either by the edge of a wheel or by the face of a grinding wheel. These machines may have reciprocating or rotating tables.

Words and Expressions

machining allowance 加工余量
removal [ri'mu:vəl] n. 除去,切削,切除
encounter [in'kauntə] v. ;n. 遇到,遇见,遭遇
tracer ['treisə] n. 追踪装置,随动装置,仿形板
merit ['merit] n. 优点,特征,价值;v. 值得,有…价值
confine [kən'fain] v. 限制在…范围内;n. 区域,范围
surface roughness 表面粗糙度
excessive [ik'sesiv] a. 过度的,极度的,非常的
centerless ['sentəlis] a. 无心的,没有心轴的
cylindrical [si'lindrikəl] a. 圆柱的,圆柱形的
taper ['teipə] n. 圆锥,锥体
gage [geidʒ] n. 量具,测量仪表,标准,限度
concentricity [ˌkɔnsen'trisiti] n. 同心,同心度
cylinder ['silində] n. 圆柱,柱体,气缸,油缸
set up 装夹,安装
grinder 磨床
pitch diameter 螺纹中径

19. Motors and Drivers

All of the mechanisms will require some type of driver to provide the input motion and energy. There are many possibilities. If the design requires a continuous rotary input motion, such as for a slider-crank or a cam-follower, then a motor is the logical choice.

Motors come in a wide variety of types. The most common energy source for a motor is electricity, but compressed air and pressurized hydraulic fluid are also used to power air and hydraulic motors. Gasoline or diesel engines are another possibility. Electric motors are made in several designs, among which are AC, DC, servo, and stepping.

AC and DC refer to *alternating current* and *direct current* respectively. AC is typically supplied by the power companies and, in the U. S., will be alternating at 60 Hertz, at about ±110 Volts or ±220 Volts.

AC motors are the most inexpensive solution to continuous rotary motion, and there are a variety of *torque-speed* curves to suit various load applications. They are limited to a few speeds, which are a function of the 60 Hz AC line frequency. The most common AC motor *no load speeds* are 1725 and 3450 revolutions per minute (RPM), which represent some slippage from the more expensive synchronous AC motor speeds of 1800 and 3600 rpm. If other output speeds are needed, a gearbox speed reducer is attached to the motor's output shaft.

DC motors are made in different electrical configurations, which provide differing *torque-speed* characteristics. The *torque-speed* curve of a motor describes how it will respond to an applied load.

Figure 19.1a shows such a curve for a permanent magnet (PM) DC motor. Note that the torque varies greatly with speed, ranging from maximum torque at zero speed to zero torque at maximum speed. This relationship comes from the fact that $Power = Torque \times Angular\ Velocity$. Since the power available from the motor is limited, an increase in torque requires a decrease in angular velocity and vice versa.

Figure 19.1b shows a family of load lines superposed on the *torque-speed* curve of the motor. These load lines represent a changing load applied to a mechanism, which the motor must supply. The problem comes from the fact that *as the required load torque increases, the motor must reduce speed to supply it.* Thus the input speed will vary in response to load variations. If constant speed is desired, this is unacceptable.

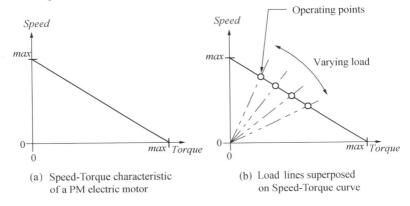

(a) Speed-Torque characteristic of a PM electric motor

(b) Load lines superposed on Speed-Torque curve

Figure 19.1 DC permanent magnet (PM) motor Speed-Torque characteristic

One possible solution is to use a speed-controlled DC motor which contains circuitry that increases and decreases the current to the motor in the face of changing load to try to maintain constant speed. These DC motors will run from an AC source since the controller converts AC to DC. The cost of this solution is high, however. Another solution is to provide a flywheel on the input shaft, which will store kinetic energy and help smooth out the speed variations introduced by the load variations.

These AC and DC motors are designed to provide continuous rotary output. Though they can be stalled against a load, they will not tolerate a full-current, zero-velocity stall for more than a few minutes without overheating.

Servomotors are fast response, closed-loop controlled motors capable of providing a programmed function of acceleration or velocity, as well as of holding a fixed position against a load. Closed loop means that *sensors on the output device being moved feed back*

information on its position, velocity, and acceleration. Circuitry in the motor controller responds to the fed back information by reducing or increasing (or reversing) the current flow to the motor. Thus precise positioning of the output device is possible, as is control of the speed and shape of its time response to changes in load or input commands. These are very expensive devices and are commonly used in applications such as moving the flight control surfaces in aircraft and guided missiles. These are relatively small and have lower power and torque capacity compared to larger AC/DC motors.

Stepping motors are designed to position an output device. Unlike servomotors, these are open loop, meaning they *receive no feedback as to whether the output device has responded as requested.* Their internal construction consists of a number of magnetic strips arranged around the circumference of both the rotor and stator. When energized, the rotor will move one step, to the next magnet, for each pulse received. Thus, these are intermittent motion devices and do not provide continuous rotary motion like other motors. The number of magnetic strips determines their resolution (typically a few degrees per step). They are relatively small compared to AC/DC motors and have low torque capacity. They are moderately expensive and require special controllers.

Air and hydraulic motors have more limited application than electric motors, simply because they require the availability of a compressed air or hydraulic source. Both of these devices are less energy efficient than direct electric to mechanical conversion of electric motors, because of the losses associated with the conversion of the energy first from chemical or electrical to fluid pressure and then to mechanical form. Every energy conversion involves some losses. Air motors find widest application in factories and shops, where high-pressure compressed air is available for other reasons. A common example is the air impact wrench used in automotive repair shop. Although individual air motors and air cylinders are relatively cheap, these pneumatic systems are quite expensive when the cost of all the ancillary equipment is included. Hydraulic motors (see Fig. 19.2) are most often found within machines or systems such as construction equipment (cranes), aircraft, and ships, where high-pressure hydraulic fluid is provided for many purposes. Hydraulic systems are very expensive when the cost of all the ancillary equipment is included.

Figure 19.2　Hydraulic motors

Air and hydraulic cylinders are linear actuators (piston in cylinder) which provide a limited stroke, straight-line output from a pressurized fluid flow input of either compressed air or hydraulic fluid (oil). They are the method of choice if you need a linear motion input. However, they share the same high cost, low efficiency, and complication factors as listed under their motor equivalents above.

Another problem is that of control. Most motors will tend to run at a constant speed. A linear actuator, when subjected to a constant pressure fluid source, typical of most compressors, will respond with more nearly constant acceleration, which means its velocity will increase linearly with time. This can result in severe impact loads on the driven mechanism when the actuator comes to the end of its stroke at maximum velocity. Servovalve control of the fluid flow, to slow the actuator at the end of its stroke, is possible but is quite expensive.

Words and Expression

driver [draivə] n. 驱动装置
slider ['slaidə] n. 滑块
pressurize ['preʃəraiz] v. 增压,对…加压,产生压力
servo ['sə:vəu] n. 伺服机构,伺服电动机,伺服传动装置
stepping ['stepiŋ] n. 步进,分级,分段
stepping motor 步进电机
alternating current and direct current 交流电和直流电
Hertz [hə:ts] n. 赫兹(频率单位)
revolutions per minute 转数/分,每分钟转数
slippage ['slipidʒ] n. 滑动量,滑移,下降,动力传递损耗,转差率
synchronous ['siŋkrənəs] a. 同步的,同时发生的
load line 负荷线,负载线
circuitry ['sə:kitri] n. 电路,线路,电路图,电路系统
kinetic [kai'netik] a. 运动的,动力的,活动的
kinetic energy 动能
stall [stɔ:l] v. 失速,停车,停止转动,发生故障
feedback ['fi:dbæk] n. 反馈,反应
stator ['steitə] n. 定子
hydraulic motor 液压马达
wrench [rentʃ] n. ;v. 扳手,拧紧
ancillary [æn'siləri] a. 辅助的,附属的;n. 辅助设备
hydraulic cylinder 液压缸
servovalve 伺服阀
actuator ['æktjueitə] n. 执行机构,致动器
impact wrench 气动扳手

20. Machine Tool Motors

Electric motors are the prime movers for most machine tool functions. They are made in a variety of types to serve three general machine tool needs: spindle power, slide drives, and auxiliary power. Most of them use 3-phase ac power supplied at 380 V.

The Basics

All electric motors use the principle that like magnetic poles repel and unlike poles attract. Current through a coil or permanent magnets creates magnetic fields. Motors deliver torque by shifting the magnetic fields within the motor, so the rotor is constantly drawn around.

Initially, all motors used direct current (dc). This current creates magnetic fields in the stator and rotor, then mechanically energizing and de-energizing stator coils cause a moving field that draws the rotor around. With alternating current (ac) motors, the current itself switches in polarity from positive to negative. Sending ac to the stator coils creates fields that draw the rotor around.

For most machine tool applications, versions of the ac asynchronous motor are used for spindle drives, while slide drives are generally synchronous.

A Little History

The design problem through the years with machine tools and motors has been how to get high torque at a variety of speeds. Initially, mechanical transmissions consisting of gears, belts, and gear/belt combinations gave speed changing capability. Up to 36 speed ranges were common at one time. But all this extra hardware is costly and needs maintenance. In the last decade, as speed requirements have risen the inaccuracy caused by vibration, which was not a problem at lower speeds, made the complex mechanical transmission unacceptable for some applications. Machine tool builders still use mechanical transmission for many applications but, because of more versatile motor speed control, three-speed transmissions are more common.

For today's operation, consider a spindle speed of 3600 rpm as low, with high speed generally 10,000 rpm and greater. At the same time, motor design and control technology have progressed dramatically. Now, thanks to computer technology, it's possible to quickly modify motor speed and torque.

Spindle Motors

A spindle is a motor-driven shaft that both positions and transmits power to a tool or holds a workpiece. Spindle motors, the major motors on a machine tool, drive the spindle shafts.

The ac asynchronous induction spindle motor is also called a squirrel cage motor (see Fig. 20.1). Alternating current—fluctuating from positive to negative in a sine wave—is sent to three stator windings causing a rotating magnetic field (see Fig. 20.2). At the same time, current induced into the rotor sets up another field. Forces set up by these two fields cause rotor rotation. To stop, a control reverses current flow.

Lower cost motors of this type control speed by varying frequency. Larger, and more precise motors, use vector control that requires a microprocessor with a math algorithm to control current accurately, both the current magnetizing the coil and the current producing torque. The motor can therefore have maximum torque at any speed.

Vector control also provides both position and current feedback in

Figure 20.1 Squirrel cage motor

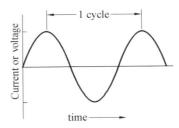

Figure 20.2 Diagram of sinusoidal alternating current

real time. Current feedback gives a good indication of what the spindle and its tool are doing (spinning in air or cutting metal). This information also determines how much time a given tool has been used, which is important for automatic tool management. In some toolchangers, spindle position control is important for toolholder orientation. It's necessary to know where the tool is for automatic toolchangers to function properly.

Another motor design now gaining acceptance as a spindle drive is the dc brushless motor with permanent magnet rotor. This motor has high torque at lower speed which is important when using larger tools. Direct current brushless motors don't develop as much heat as comparable ac motors, but are larger, more costly, and have speed limitations.

Feed Motors

Positioning motors drive the ballscrews that move the slides carrying spindles or worktables. With positioning motors, the key features are acceleration, deceleration, smoothness, and accuracy so the slide or spindle gets to where it's needed quickly. As with spindle motors, positioning motors were initially dc brush type. Then ac induction became popular, followed by hydraulic drives. Hydraulics have good acceleration and power-to-weight ratio because of low mass. Although still used in some high repeatability applications, they were generally displaced by ac servos that offer longer life, higher efficiency, and less heat generation.

Accurate Positioning

Position control is the key feature in feed motor operation. It is important to know that accurate positioning depends on feedback, or a closed-loop system. Accuracy depends on how well the system was made and if anything interfered with system motion.

With feedback or closed loop, the motor obeys the command, but has a sensor that sends back the signal, "Here's where I am." If it's not the right place, the control warns the operator or sends a corrective signal. This "conversation" between the motor and control takes place almost instantaneously.

Motors tell where they are with encoders and resolvers. An encoder is a device attached to the motor shaft. It generates a digital signal that notes how many turns or partial turns the motor shaft made. A resolver is a device on a motor that generates a sine wave as the motor turns. A controller senses the sine wave, counting both the number of waves and sine angle to establish where the motor is.

Words and Expressions

prime mover 原动力,原动机(转化自然能使其做功的机器或装置)
auxiliary [ɔ:g'ziljəri] a. 辅助的
3-phase ac 三相交流电
like magnetic pole 相同的磁极
energize ['enədʒaiz] v. 激发,给予…电压,使…带电
deenergize v. 切断,断开,释放(继电器,电磁铁等),去(解除)激励
polarity [pəu'læriti] n. 极性
asynchronous [ei'siŋkrənəs] a. 不同时的,异步的
synchronous ['siŋkrənəs] a. 同步的,同时发生的,同时出现的
inaccuracy [in'ækjurəsi] n. 误差,不准确
mechanical transmission 机械传动
grinding quill 磨床主轴
power conditioner 动力(功率)调节器
armature ['ɑ:mətjuə] n. 电枢,转子
squirrel cage motor 鼠笼式电动机
fluctuate ['flʌktjueit] v. 脉动,振荡,变化
sine wave 正弦波形
vector ['vektə] n. 矢量,向量
magnetize ['mægnitaiz] v. 使磁化
real time 实时(在数据发生的当时处理该项数据,并在所需的响应时间以内获得必要的结果。)
positioning [pə'ziʃəniŋ] n. 定位,位置控制
ballscrew ['bɔ:lskru:] n. 滚珠丝杠
worktable 工作台
induction [in'dʌkʃən] n. 感应
induction motor 感应电机
hydraulics 水力学,液压系统

activate ['æktiveit] v. ;n. 开动,启动,驱动,激发
interfere with 干涉,干扰,妨碍
encoder [in'kəudə] n. 编码器(一种能提供位置反馈和速度反馈的测量装置)
resolver [ri:'zɔlvə] n. 分解器(一种能将旋转的和线性的机械位移转换成模拟电信号的变换器)
limit switch 限位开关,极限开关
trip 行程,释放,松开,断路,断开
bump [bʌmp] v. ;n. 撞,碰撞,冲击
overlap [əuvə'læp] v. ;n. 重叠,叠加,相交,交错
operating cycle 工作循环,操作循环

21. Numerical Control

One of the most fundamental concepts in the area of advanced manufacturing technologies is numerical control (NC). Prior to the advent of NC, all machine tools were manually operated and controlled. Among the many limitations associated with manual control machine tools, perhaps none is more prominent than the limitation of operator skills. With manual control, the quality of the product is directly related to and limited to the skills of the operator. Numerical control represents the first major step away from human control of machine tools.

Numerical control means the control of machine tools and other manufacturing systems through the use of prerecorded, written symbolic instructions. Rather than operating a machine tool, an NC technician writes a program that issues operational instructions to the machine tool.

Numerical control was developed to overcome the limitation of human operators, and it has done so. Numerical control machines are more accurate than manually operated machines, they can produce parts more uniformly, they are faster, and the long-run tooling costs are lower. The development of NC led to the development of several other innovations in manufacturing technology:

1. Electrical discharge machining.
2. Laser cutting.
3. Electron beam welding.

Numerical control has also made machine tools more versatile than their manually operated predecessors. An NC machine tool can automatically produce a wide variety of parts, each involving an assortment of widely varied and complex machining processes. Numerical control has allowed manufacturers to undertake the production of products that would not have been feasible from an economic perspective using manually controlled machine tools and processes.

Like so many advanced technologies, NC was born in the laboratories of the Massachusetts Institute of Technology. The concept of NC was developed in the early 1950s with funding provided by the U. S. Air Force.

The APT (Automatically Programmed Tools) language was designed at the Servomechanism laboratory of MIT in 1956. This is a special programming language for NC that uses statements similar to English language to define the part geometry, describe the cutting tool configuration, and specify the necessary motions. The development of the APT language was a major step forward in the further development of NC technology. The original NC systems were vastly different from those used today. The machines had hardwired logic circuits. The instructional programs were written on punched paper tape (see Fig. 21.1), which was later to be replaced by magnetic plastic tape. A tape reader (see Fig. 21.2) was used to interpret the instructions written on the tape for the machine. Together, all of this represented a giant step forward in the control of machine tools. However, there were a number of problems with NC at this point in its development.

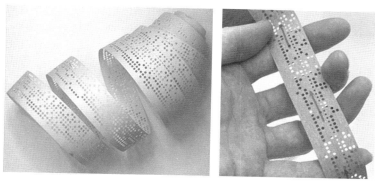

Figure 21.1 Punched paper tapes

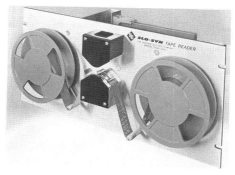

Figure 21.2 A paper tape reader

A major problem was the fragility of the punched paper tape medium. It was common for the paper tape containing the programmed instructions to break or tear during a machining process. This problem was exacerbated by the fact that each successive time a part was produced on a machine tool, the paper tape carrying the programmed instructions had to be rerun through the reader. If it was necessary to produce 100 copies of a given part, it was also necessary to run the paper tape through the reader 100 separate times. Fragile paper tapes simply could not withstand the rigors of a shop floor environment and this kind of repeated use.

This led to the development of a special magnetic plastic tape. Whereas the paper tape carried the programmed instructions as a series of holes punched in the tape, the plastic tape carried the instructions as a series of magnetic dots. The plastic tape was much stronger than the paper tape, which solved the problem of frequent tearing and breakage. However, it still left two other problems.

The most important of these was that it was difficult or impossible to change the instructions entered on the tape. To make even the most minor adjustments in a program of instructions, it was necessary to interrupt machining operations and make a new tape. It was also still necessary to run the tape through the reader as many times as there were parts to be produced. Fortunately, computer technology became a reality and soon solved the problems of NC associated with punched paper and plastic tape.

The development of a concept known as direct numerical control (DNC) solved the paper and plastic tape problems associated with numerical control by simply eliminating tape as the medium for carrying the programmed instructions. In direct numerical control, machine tools are tied, via a data transmission link, to a host computer. Programs for operating the machine tools are stored in the host computer and fed to the machine tool as needed via the data transmission linkage. Direct numerical control represented a major step forward over punched tape and plastic tape. However, it is subject to the same limitations as all technologies that depend on a host computer. When the host computer goes down, the machine tools also experience downtime. This problem led to the development of computer numerical control.

The development of the microprocessor allowed for the development of programmable logic controllers (PLCs) and microcomputers. These two technologies allowed for the development of computer numerical control (CNC). With CNC, each machine tool has a PLC (see Fig. 21.3) or a microcomputer that serves the same purpose. This allows programs to be input and stored at each individual machine tool. It also allows programs to be developed off-line and downloaded at the individual machine tool. CNC solved the problems associated with downtime of the host computer, but it introduced another problem known as data management. The same program might be loaded on ten different microcomputers with no communication among them. This problem is in the process of being solved by local area networks that connect microcomputers for better data management.

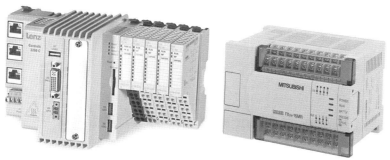

Figure 21.3 Programmable logic controllers

Words and Expressions

prerecord ['priːriˈkɔːd] v. 事先录制
symbolic instruction 符号指令

operational instruction 操作指令
electrical discharge machining 电火花加工
laser cutting 激光切割
electron beam welding 电子束焊接
predecessor ['priːdisesə] n. 前辈，前任，(被取代的)原有事物
assortment [ə'sɔːtmənt] n. 分类，种类
an assortment of 各式各样的
hardwired logic 硬连线逻辑
circuit ['səːkit] n. 电路
punched paper tape 穿孔纸带
tape reader 带阅读器，带读取器
programmed instruction 程序指令
shop floor 工作场所，车间
direct numerical control 直接数字控制
host computer 主计算机(在多机系统中起主要作用或控制作用的计算机)
successive [sək'sesiv] a. 连续的，连贯的，相继的
programmable logic controller 可编程序逻辑控制器
off-line 脱机，离线

22. Choice of Manual or CNC Machine Tools

  The CNC machine offers enormous flexibility in producing short-run parts, but may not be as economical as a dedicated non-CNC machine with appropriate automation in volume production. Oddly enough, it may not be the most economical choice for one-off toolroom or repair shop applications either.

  The emphasis in both cases is on the word "economical", because capacity for the CNC machine can be expected to cost significantly more than a manual counterpart. So, what do you get for the extra money?

Flexibility is the strong suit of CNC turning machines, with accuracy and repeatability not far behind. We may not like to admit it, but today's electro-mechanical systems are every bit the equal of the most skilled machinist who ever lived, and they never get tired. In a world where quality is increasingly a statistical proposition, it's hard to argue with the uniformity of a computer-controlled process.

On the other hand, if all the machine is required to do is turn out cylindrical repair parts on an "as-needed" basis, you might want to seriously consider a manual machine. Not only will it cost less, but chances are that even a moderately skilled machinist will be able to make simple parts on it just about as quickly as he can program and run the CNC. The advantage may shift to the CNC, however, if the same part has to be made frequently, if the parts are complex, or, as may be the case more and more often in the future, if there is nobody available who knows how to operate a manual machine.

All things considered, a CNC machine is probably the better choice in a situation in which the economics are anywhere close to equal. And where flexibility, accuracy, and repeatability are the key requirements, it's clearly the right answer.

CNC or manual? This is an easy choice when you're evaluating turning or milling machines, and many of the same considerations come into play with grinders, but often for very different reasons. For example, consider the characteristics of the process.

One of the main advantages of CNC in a milling or turning machine is the ability to easily program complex motions of the tool to produce required surface features on the workpiece. In a grinder, all you have to do is dress the required features into the wheel and they're automatically transferred to the workpiece. So, why would you want to pay the higher price of a CNC grinder?

The answer is for control, but not necessarily the kind of direct control over workpiece features you get in milling or turning. In grinding, the CNC's advantage is that it lets you control the process, and that directly impacts the dimensional, geometric, and surface roughness of the workpiece. Perhaps the point is best illustrated with a fairly sophisticated example.

Automotive camshaft lobes traditionally have been ground between centers in a plunge-type operation in which wheel feed is controlled by a cam synchronized to workpiece rotation. While this arrangement was able to meet historically acceptable standards for size, finish, and lobe geometry while delivering adequate productivity, it has proven increasingly unable to achieve the more rigid quality demands of today's higher power-density engines at acceptable production rates.

One answer has been to convert the process to CNC control. In these machines, the spindle, wheel-feed, and workhead are all servo-driven under CNC control. What that means in practice is that grinding conditions can be adjusted for optimum performance at every point on the cam lobe surface by changing the wheel and/or workpiece speed to achieve the most efficient conditions for grinding that exact point.

Where the cam-type grinder necessarily incorporated a whole series of compromises into the operation, the CNC machine requires virtually none. The result is a more efficient, more precise, and more productive process.

This is a good example of one reason why grinder manufacturers are more likely than turning or milling machine producers to design and build their own CNC's. The controls have to be uncommonly fast and powerful, and they have to deal with specialized conditions that aren't normally present in non-grinding machine tools.

A CNC grinder will also be able to handle a CNC dresser which can improve both size control and wheel life when integrated into a comprehensive control system. The CNC dresser, essentially a small diamond-tooled lathe, makes it easy to true the wheel for cylindrical or flat surface grinding, and to produce simple or complex contours for form grinding.

Of course, the CNC also makes it easier to control relatively simple workpiece attributes like size, where the latest generation of in-process gages work synergistically with the control to hold both size and geometry to tolerances achievable only in the laboratory just a few years ago.

Finally, a CNC grinder is probably easier to set-up and use for short-run applications. All things considered, however, its major advantage lies in the precision control it offers over the process itself.

Words and Expressions

enormous [i'nɔːməs] *a.* 巨大的
short-run 短期的, 少量生产, 短期生产
dedicated ['dedikeitid] *a.* 专用的
volume production 批量生产, 成批生产
oddly enough 说也奇怪
one-off 一次性的, 单件的; 一次性事物, 单件生产
toolroom [tuːlrum] *n.* 工具室, 工具车间
counterpart ['kauntəpɑːt] *n.* 相似物, 对应物, 另一个具有相同功能或特
　　　　　　　　　　　　　　色的人或物
one's strong suit 优点, 长处
electro-mechanical 机电的
proposition [prɔpə'ziʃən] *n.* 提议, 主张, 定理, 问题
uniformity [juːni'fɔːmiti] *n.* 同样, 一致, 均匀, 一致性, 均匀性
turn out 生产, 制造

machinist [məˈʃiːnist] n. 机械工人,机工
dress 对砂轮进行修锐和整形
camshaft [ˈkæmʃɑːft] n. (机械)凸轮轴
lobe [ləub] n. 凸起,凸起部
plunge-type 横向进给(磨削)方式,切入式(磨削)
cam [kæm] n. 凸轮
synchronize [ˈsiŋkrənaiz] v. 同步
high power density 大功率密度
work head 转盘,工作台
compromise [ˈkɔmprəmaiz] n. ;v. 折衷,平衡,综合考虑
dresser 修整器(为了获得一定的几何形状和锐利磨削刃,用金刚石等工具
    对砂轮表面进行修整的装置)
to true the wheel 对砂轮进行整形(产生确定的几何形状)
form grinding 成形磨削
in-process 加工过程中的
gage = gauge [geidʒ] n. 测量仪表,传感器,检验仪器
synergistic [sinəˈdʒistik] a. 协同的,合作的,互相作用的

23. Hard-Part Machining with Ceramic Inserts

Hard-part machining (HPM), the machining of extremely hard workpieces using ceramic inserts (see Fig. 23.1), permits metalworking plants to cut costs and improve product quality at a modest startup cost. It can be done successfully by any skilled turning-machine operator, whether in a machine-shop environment involving small or prototype runs or in a high-volume-production operation. HPM can be performed on relatively inexpensive machine tools as long as the machine is sufficiently rigid, and high-quality, uniform ceramic inserts are used.

Figure 23.1 Ceramic inserts and indexable turning tools

Machining hardened parts involves materials registering 55 to 65 HRC and tensile strengths to 2400 MPa (350000 psi). Before ceramic inserts were developed, vitrified-bond alumina grinding wheels were used to obtain the desired surface roughness. Next, a process for machining hardened rolls using ceramic and other composite materials was developed.

Hard-part machining was then developed for the automobile industry. Hot-pressed alumina inserts were used for this application. Since then, other tool materials have been developed, resulting in overall acceptance of the HPM process.

Cutting Forces: a Critical Factor

With HPM, two major controlling factors must be considered: selection of the proper speed and proper tool angles. The speed depends on the hardness of the material; in general, as the hardness increases, the speed decreases. The determination of the tool angles depends on a variety of parameters: part hardness, the machine tool selected, the part's surface roughness and tolerance required, and the part material.

In addition to these parameters, it is important to consider the cutting forces. There are three forces generated in every metal-removal process: tangential force, generated by the part rotation; radial force, generated by the resistance of the workpiece material to depth of cut; and, lastly, longitudinal force, generated by the feed rate applied. These forces are 30% to 80% greater than in "soft" machining processes. For example, when comparing preheat-treated to heat-treated steel with a hardness of 62 HRC, the longitudinal force increases from 30% to 50%, the tangential force increases from 30% to 40%, and the radial force increases from 70% to 100%. Therefore, the machine tool must be able to handle the increased cutting forces, especially in the radial direction. However, these increases are small enough to have little effect on small- to medium-volume operations and will require purchase of no new equipment.

Advantages over Conventional Grinding

For grinding cylindrical applications, both the wheel and the workpiece must rotate. Moreover, the wheel rotates rapidly while the workpiece rotates slowly. If the rotating members are imperfectly concentric, the combination of imperfections and rotational-speed differential produces lobing. A geometric out-of-round pattern on the workpiece is produced, which can affect the end-product performance. With HPM, on the other hand, either the workpiece or cutting tool is rotated, not both. Therefore, the machined surface will be as accurate as the machine-tool spindle and the longitudinal direction of the machine tool relative to the center line of the machine.

Another disadvantage with grinding is the generation of tremendous surface heat at the point of contact between the grinding wheel and the workpiece. Even when flood coolant is properly applied, workpiece surface stress risers and heat checks can occur, which can lead to premature failure of the ground part in service. With ceramic HPM, less heat is generated, and if properly applied, the heat that is generated will be carried away with the brittle material removed. Thus, the finished parts are produced without stress risers or heat checks.

Another major advantage of HPM is that conventional turning machines can be used with workpieces as hard as 65 HRC using commercially available ceramic inserts. Savings occur in two areas, processing and capital investment. In processing, the machining, setup, and tool-changing time are significantly reduced. Grinding-wheel changing, on the other hand, is time-consuming. Guards must be removed, along with the spindle locking nuts, the worn wheel must be changed, and the new wheel balanced and dressed. Wheel changing can take as much as 100 times longer than changing ceramic inserts, which require only simple indexing or replacement in the holder.

Equipment also is less expensive. A turning machine costs significantly less than a production grinder to do comparable work. As already mentioned, setup is easier and quicker. Turning machines also are simpler in construction—there are no reciprocating slides to wear, maintain, or replace—for easier maintenance. However, the strength and rigidity of every component in the machine must be adequate to handle the additional cutting forces.

In addition, HPM usually requires no coolant. Dry cutting eliminates not only coolant costs, but also the expense of related housekeeping. Coolant mist can permeate the plant atmosphere and infiltrate machine controls, and coolant residue can carry grinding-wheel media into material-handling equipment. Spillage and leaks can cause slippery, hazardous areas. Disposal of chips during HPM poses no problem; chips in the machine collection tray have the consistency of dry, brittle steel wool, which disintegrates into very fine powder and can be readily compacted to a small volume for safe, easy disposal.

Words and Expressions

hard [hɑːd] *a.* 硬的,淬硬的
hard-part machining (HPM) 硬态切削
insert [inˈsəːt] *n.* 刀片
ceramic insert 陶瓷刀片
modest [ˈmɔdist] *a.* 适度的,合适的
startup = start-up 启动,开始工件,运转
turning machine 车床
high volume production 大批量生产
indexable turning tool 可转位车刀
register [ˈredʒistə] *n.* 记录; *v.* 显示出,测量
PSI = pounds per square inch 磅/平方英寸
vitrified-bond 陶瓷结合剂
alumina grinding wheel 氧化铝砂轮
hot-pressed alumina insert 热压氧化铝刀片
critical factor 关键因素
tool angles 刀具角度
hardness [ˈhɑːdnis] *n.* 硬度
metal removal 金属切削
tangential [tænˈdʒənʃəl] *a.* 切线的,切向的
radial [ˈreidiəl] *a.* 径向的; *n.* 径向
longitudinal [lɔŋdʒiˈtjuːdinəl] *a.* 纵向的,长度的,轴向的
feed rate 进给速度
preheat [priːˈhiːt] *n.* 预先加热
heat-treated 热处理的,热处理过的
imperfect [imˈpəːfiktli] *ad.* 有缺点地,不完善地
concentric [kənˈsentrik] *a.* 同中心的,同轴的
lobing [ˈləubiŋ] *n.* (圆柱的)凸角

out of round 不(很)圆,失圆
end product 最后结果,最终产品,成品
heat check 热裂纹,热致裂纹
premature ['premə'tjuə] a. 过早的,未成熟的,不到期的
ground part 经过磨削加工的零件
capital investment 资本投资,基本建设投资
guard n. 保护(器,装置),防护(器,罩,装置)
nut [nʌt] n. 螺帽,螺母
housekeeping 辅助工作,服务性工作
permeate ['pə:mieit] v. 渗入,渗透,弥漫,充满
infiltrate ['infiltreit] v. 渗透,渗进
material-handling 物料输送
spillage ['spilidʒ] n. 溢出[溅出,倒出](的物质),泄漏
slippery ['slipəri] a. 滑的,打滑的
hazardous ['hæzədəs] a. 危险的
disposal [di'spəuzl] n. 处理,处置
disintegrate [dis'intigreit] v. (使)分散[离,解,化],粉碎,切碎

24. Nontraditional Manufacturing Processes

The human race has distinguished itself from all other forms of life by using tools and intelligence to create items that serve to make life easier and more enjoyable. Through the centuries, both the tools and the energy sources to power these tools have evolved to meet the increasing sophistication and complexity of mankind's ideas.

In their earliest forms, tools primarily consisted of stone instruments. Considering the relative simplicity of the items being made and the materials being shaped, stone was adequate. When iron tools were invented, durable metals and more sophisticated articles could be produced. The twentieth century has seen the creation of products made from the most durable and, consequently, the most difficult-to-machine materials in history. In an effort to meet the manufacturing challenges created by these materials, tools have now evolved to include materials such as alloy steel, carbide, diamond, and ceramics.

A similar evolution has taken place with the methods used to power our tools. Initially, tools were powered by muscles; either human or animal. However as the powers of water, wind, steam, and electricity were harnessed, mankind was able to further extended manufacturing capabilities with new machines, greater accuracy, and faster machining rates.

Every time new tools, tool materials, and power sources are utilized, the efficiency and capabilities of manufacturers are greatly enhanced. However as old problems are solved, new problems and challenges arise so that the manufacturers of today are faced with tough questions such as the following: how do you drill a 2-mm diameter hole 670-mm deep without experiencing taper or runout? Is there a way to efficiently deburr passageways inside complex castings and guarantee 100% that no burrs were missed? Is there a welding process that can eliminate the thermal damage now occurring to my product?

Since the 1940s, a revolution in manufacturing has been taking place that once again allows manufacturers to meet the demands imposed by increasingly sophisticated designs and durable, but in many cases nearly unmachinable, materials. This manufacturing revolution is now, as it has been in the past, centered on the use of new tools and new forms of energy. The result has been the introduction of new manufacturing processes used for material removal, forming, and joining, known today as nontraditional manufacturing processes.

MACHINE TOOLS AND MACHINING 121

The conventional manufacturing processes in use today for material removal primarily rely on electric motors and hard tool materials to perform tasks such as sawing, drilling, and broaching. Conventional forming operations are performed with the energy from electric motors, hydraulics, and gravity. Likewise, material joining is conventionally accomplished with thermal energy sources such as burning gases and electric arcs.

In contrast, nontraditional manufacturing processes harness energy sources considered unconventional by yesterday's standards. Material removal can now be accomplished with electrochemical reactions, high-temperature plasmas, and high-velocity jets of liquids and abrasives (see Fig. 24.1). Materials that in the past have been extremely difficult to form, are now formed with magnetic fields, explosives, and the shock waves from powerful electric sparks. Material-joining capabilities have been expanded with the use of high-frequency sound waves and beams of electrons.

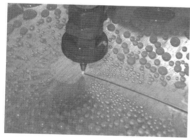 

Figure 24.1 Water jet machining process

In the past 50 years, over 20 different nontraditional manufacturing processes have been invented and successfully implemented into production. The reason there are such a large number of nontraditional processes is the same reason there are such a large number of conventional processes; each process has its own characteristic attributes and limitations, hence no one process is best for all manufacturing situations.

For example, nontraditional process are sometimes applied to increase productivity either by reducing the number of overall manufacturing operations required to produce a product or by performing operations faster than the previously used method.

In other cases, nontraditional processes are used to reduce the number of rejects experienced by the old manufacturing method by increasing repeatability, reducing in-process breakage of fragile workpieces, or by minimizing detrimental effects on workpiece properties.

Because of the aforementioned attributes, nontraditional manufacturing processes have experienced steady growth since their introduction. An increasing growth rate for these processes in the future is assured for the following reasons:

1. Currently, nontraditional processes possess virtually unlimited capabilities when compared with conventional processes, except for volumetric material removal rates. Great advances have been made in the past few years in increasing the removal rates of some of these processes, and there is no reason to believe that this trend will not continue into the future.

2. Approximately one-half of the nontraditional manufacturing processes are available with computer control of the process parameters. The use of computers lends simplicity to processes that people may be unfamiliar with, and thereby accelerates acceptance. Additionally, computer control assures reliability and repeatability, which also accelerates acceptance and implementation.

3. Most nontraditional processes are capable of being adaptively-controlled through the use of vision systems, laser gages, and other in-process inspection techniques. If, for example, the in-process inspection system determines that the size of holes being produced in a

product are becoming smaller, the size can be modified without changing hard tools, such as drills.

4. The implementation of nontraditional manufacturing processes will continue to increase as manufacturing engineers, product designers, and metallurgical engineers become increasingly aware of the unique capabilities and benefits that nontraditional manufacturing processes provide.

Words and Expressions

durable ['djuərəbl] *a.* 耐用的,耐久的;*n.* 耐用的物品
carbide ['kɑːbaid] *n.* 碳化物,硬质合金
difficult-to-machine material 难加工材料
taper ['teipə] *n.* 锥度,圆锥
runout ['rʌn'aut] *n.* 偏斜,径向跳动
deburr [di'bəː] *v.* 去毛刺
passageway ['pæsidʒ'wei] *n.* 通道,通路
burr [bəː] *n.* 毛刺
sawing ['sɔːiŋ] *n.* 锯,锯开,锯切
unconventional ['ʌnkən'venʃənl] *a.* 非传统的,不是常规的
electrochemical [iˈlektrəuˈkemikl] *a.* 电化学的
plasmas ['plæzmə] *n.* 等离子
reject [ri'dʒekt] *n.* 等外品,不合格品,废品
in-process *a.* (加工,处理)过程中的
breakage ['breikidʒ] *n.* 破损,断裂,损坏
fragile ['frædʒail] *a.* 易碎的,易毁坏的
minimize ['minimaiz] *v.* 使…成为最小,最小化
detrimental [ˌdetri'mentəl] *a.* 有害的,不利的
aforementioned [ə'fɔː'menʃənd] *a.* 上述的,前面提到
adaptive [ə'dæptiv] *a.* 适合的,适应的,自适应的

adaptive control 自适应控制

25. Design Considerations for NC Machine Tools

The design of NC machine tools requires a multifaceted approach which considers both control system response characteristics and the mechanical characteristics of the machine structural members. Because the NC machine is a complex system, an integrated approach must be taken and an individual component must be designed so that it will conform to the criteria defined for the entire system. The most accurate control system is of limited value if the mechanical actuation components do not adequately respond to commands.

Once the performance requirements of the machine tool are determined, the overall NC system can be designed. General parameters, such as: machine function; workpiece size, type and material; accuracy and surface roughness; power consumption; control features; and auxiliary functions, all lead to the proper selection of an integrated control system and machine configurations.

The benefits of an NC machine are directly related to its accuracy, speed, and automaticity. However, it is these same characteristics that create design problems. A manually operated machine tool has an intelligent source for error compensation. That is, a machinist who has worked with a particular machine over a long period of time learns its characteristics. Under a given set of operating conditions, experience has taught him that one or two thousandths error results from machine member deflection. The knowledgeable operator can then compensate.

The NC machine tool can only compensate for an error that is detected and communicated to the control unit. Structural compliance, vibration, and other aspects of mechanical design cannot be easily or economically monitored. For this reason, an NC machine is designed to be stronger, stiffer, and to perform to a more accurate standard than its conventional counterpart.

NC and conventional equipment achieve different dynamic performance. The NC machine elements must be capable of performing accurately under high accelerations and decelerations. Because leadscrews and gearing are an integral part of the closed loop system, mechanical inaccuracy such as backlash and deflection, which tend to produce instability, must be reduced to levels not required for manually operated machines.

The control system of an NC machine tool is required to respond accurately under both steady state and dynamic operating conditions. A general criterion for the design of machine structures for numerical control is to provide adequate static stiffness with the best stiffness to weight ratio for a broad range of loading conditions. The dynamic response of the machine is directly related to the stiffness to weight ratio. Stiff machines with low mass have a rapid dynamic response.

Other aspects related to the mechanical nature of the NC machine (e. g. , friction, backlash, slip-stick) can adversely affect the machine performance if they are not compensated or minimized. Vibration, resulting from tool chatter, can have a devastating effect on the quality of surface roughness and accuracy. In extreme conditions it may lead to machine tool damage.

Dry friction is not desirable and is minimized in NC machines. Dry friction leads to a phenomenon called stick-slip.

Under conditions of low normal stress found on NC machine slides, Amonton's law is applicable. Thus,

$$F = \mu N$$

where F is the friction force resulting from a coefficient of friction μ

and normal force N.

The initial large coefficient of friction between the table and slides requires a considerable frictional force to initiate movement, that is, to overcome the *sticking* of the components. Once motion begins, μ falls and the drive force necessary to sustain movement decreases rapidly, causing the machine table to move beyond the point where driving force equals the sliding friction force. This slipping action results from the inertia of the table and the spring effect due to a rapid release of energy. Although the stick-slip phenomenon only occurs at low feedrates, it is detrimental to system performance and should be minimized.

Antifriction bearings are used to reduce both the coefficient of friction between the machine slide and ways, and the phenomenon of stick-silp. Two basic types of bearings are used—rolling bearings and hydrostatic bearings. Both types result in low coefficients of friction and have inherent advantages and disadvantages. The choice of a particular bearing depends upon normal force level, cost, and desired accuracy.

To illustrate the use of antifriction devices in NC machine tools, consider the hydrostatic bearing schematically illustrated in Fig. 25.1a. Hydraulic fluid from a constant pressure source flows through the bearing and is discharged over the bearing area in such a manner that a pressure distribution of the form shown in Fig. 25.1b results. The pressure is balanced by a downward load which enables the bearing to ride on a layer of fluid, H.

The advantages of the hydrostatic bearing include the elimination of the stick-slip phenomenon and a reduction in the coefficient of friction to approximately 10^{-5}. Because the surfaces are separated, wear is negligible, and errors in the accuracy of the ways are averaged, enabling less costly machined surfaces to be used.

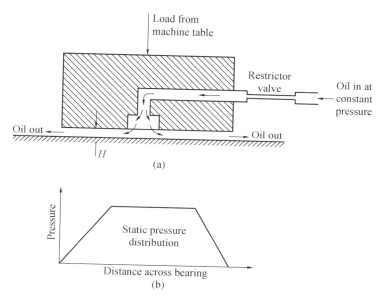

Figure 25.1 (a) Hydrostatic bearing schematic, (b) Pressure distribution

Hydrostatic bearings are limited to low normal stress applications because of film breakdown.

To reduce deflection in machine tool members, the stiffness of each member must be sufficient to counteract applied loads. Therefore, machine members tend to have large mass and inertia. Mass not contributing to stiffness is eliminated in well designed machines.

Words and Expressions

facet ['fæsit] n. 方面,小平面
multifaceted a. 多方面的,多层次的
structural member 结构件,构件
integrated approach 综合方法
machine function 机器功能,机器作用

machine configuration 机器配置
integrated control system 综合控制系统
automaticity [ɔː'tɔmətisəti] n. 自动性,自动化程度
operating condition 运行条件,工作状况,操作条件
compliance [kəm'plains] n. 依从,顺从,柔度
dynamic performance 动态特性,动力特性
lead screw 导杆,丝杠
steady state 稳定状态
loading condition 载荷条件,负荷条件
backlash ['bæklæʃ] n. 间隙,齿间隙
slip-stick 滑动面粘附现象
chatter ['tʃætə] n. 振动,颤震现象
devastating ['devəsteitiŋ] a. 破坏性的
Amonton's law 阿蒙登摩擦定律
coefficient of friction 摩擦系数
normal force 法向力
slide 滑板(顶部与有关零、部件连接,底面具有导轨,可在相配合的零、部件上移动的部件)
sticking ['stikiŋ] a. 粘的,有粘性的
instability [instə'biliti] n. 不稳定性
stick-slip 蠕动,爬行
antifriction bearing 减摩轴承,滚动轴承
hydrostatic bearing 流体静压轴承
schematically [ski'mætikli] ad. 用示意图,示意地
oil in 进油,进油口
oil out 出油,出油口
restrictor valve 节流阀
counteract [kauntə'rækt] v. 抵消,中和,阻碍

26. Laser-Assisted Machining and Cryogenic Machining Technique

To meet the demand for production and precision, researchers and equipment builders are looking outside the bounds of conventional milling, drilling, turning, and grinding. New cutting and machining processes continue to emerge along with methods for modifying the conventional techniques.

Laser-assisted machining. Laser-assisted machining is a process that has long been researched and is now ready to come out of the lab. In this operation, a laser beam is projected onto the part through fiber optics or some other optical-beam delivery unit, just ahead of the tool (as shown in Fig. 26.1). During a cut, laser power is varied to match the profile being cut. CO_2 lasers are usually used with power levels from 200 to 500 W. The laser-induced heat softens the workpiece and makes it easier to cut.

Figure 26.1 Laser-assisted machining process

This process can be used for very difficult-to-machine materials such as superalloy or ceramics. For example, with such process you can cut through ceramics like butter. It also offers a good surface roughness, usually 0.5 μm in Ra or lower.

Because the heating laser beam is tightly focused, heating is localized around the actual cutting zone. Heat is carried off in the chips and there are no changes in the physical properties of the material cut due to heat.

Cryogenic machining technique. An unique use of liquid nitrogen is offering a modification of traditional turning to improve manufacturing operations. Using $-195\,^\circ\!\mathrm{C}$ liquid nitrogen as a coolant as a means to lower the heat in the cutting zone offers a number of advantages. The cryogenic coolant system delivers a jet of $-195\,^\circ\!\mathrm{C}$ liquid nitrogen directly to the insert during turning operations (as shown in Fig. 26.2). This raises insert hardness, which significantly reduces the thermal softening effect that an insert may experience as a result of hard turning's inherent high cutting temperatures. The steep temperature gradient between the chip/tool interface and insert body also helps remove heat from the cutting zone.

Figure 26.2 Cryogenic machining process

Cubic boron nitride (CBN) and polycrystalline cubic boron nitride (PCBN) inserts have traditionally been the tools of choice for hard-turning applications.

But, they are considered too costly for many operations. The cryogenic machining technique also allows greater use of low-cost ceramic inserts for hard-turning operations.

CBN and PCBN ceramic inserts tend to wear unevenly and are prone to fracturing when hard turning dry or with water or oil-based coolants. Increased fracture toughness resulting from low-temperature liquid nitrogen cooling provides more predictable, gradual flank wear for ceramic inserts, as well as increasing cutting speeds up to 200%. This predictable flank wear also allows alumina ceramic inserts to be used in critical finishing operations.

The liquid nitrogen may be stored in a small, dedicated cylinder near a machine, or in a supply tank that would serve multiple machines. Programming is similar to a traditional coolant delivery system. A flexible liquid nitrogen line attaches to a lathe's turret via a rotational coupling. This line feeds a delivery nozzle clamped to the tool, which directs the liquid nitrogen to the insert tip.

Nitrogen is a safe, noncombustible, and noncorrosive gas. It quickly evaporates after contact with the insert, and returns back into the atmosphere, leaving no residue to contaminate the workpiece, chips, machine tool, or operator, and it also eliminates disposal costs. This is particularly helpful for porous powder metal parts, which often require subsequent part cleaning operations to remove coolant residue.

Now these two processes have become commercially available.

Words and Expressions

laser-assisted machining 激光辅助切削
difficult-to-machine material 难加工材料
superalloy [ˌsjuːpəˈælɔi] n. 高温合金,超耐热合金

turning ['tə:niŋ] n. 车削
cutting zone 切削区
cryogenic [ˌkraiəu'dʒenik] a. 低温学的
insert 刀片
cubic boron nitride (CBN) 立方氮化硼
polycrystalline cubic boron nitride (PCBN) 聚晶立方氮化硼
hard-turning 硬态车削
hard turning dry 硬态干式车削
fracture toughness 断裂韧度
flank wear 后刀面磨损
alumina ceramic insert 氧化铝陶瓷刀片
finishing operation 精加工
lathe [leið] n. 车床; vt. 用车床加工
turret ['tʌrit] n. 转塔刀架
coupling ['kʌpliŋ] n. 联轴器,管接头
residue ['rezidju:] n. 残留物
chip 切屑
disposal cost 处理费用
powder metal part 粉末冶金零件

3   QUALITY AND PRODUCTIVITY

27. Quality in Manufacturing

Machine tool quality

A recent survey by the Society of Manufacturing Engineers reveals that U. S. manufactures, by two to one, rate foreign-built machine tools as being more trouble-free than U. S.-made production equipment. Conversely, American manufacturers rate service of equipment by U. S. vendors as better than that provided by offshore supplies, by a two to one ratio. U. S. firms also respond faster with service, by three to one, and parts availability is better, by eight to one.

The survey presents an overview of the U. S. machine tool industry based on four factors which SME terms critical to quality performance in machine tools: installation and startup, consistent operation at required production rates, trouble-free operation, and service support. The survey was sent to 2,000 engineers and managers in five basic industries; metal products, machinery/computer equipment, electrical/electronic equipment, transportation equipment, and instruments/optical goods.

Some of the respondents commented that U. S. equipment manufacturers are playing catch-up and only recently are improving productivity, quality, and price. Others believe that systems integration is poor from U. S. suppliers compared to the Europeans. Fortunately, quality improvement programs are underway at a growing number of machine tool companies in an effort to meet the foreign competitive challenge. One company claims that more than 70% of their employeess have at least 20 hours of training in quality.

Quality control at an automotive supplier

For those component manufacturers who may be supplying parts to the automotive industry in the future, Associated Spring (Bristol, CT) may be a good example to follow. Associated Spring (AS) is now Ford's predominant valve spring supplier and the largest supplier of valve springs to the North American automotive industry. However, AS had to redouble their quality efforts to win back the Ford account, which it lost in 1982.

AS has established a quality program that involves both internal and external (suppliers) standards. The guiding philosophy behind this program begins by saying that AS exists to serve their customers, which requires building partnerships with customers and suppliers. The company must also support this philosophy with a commitment to investing in training, re-educating all employees, and equipment and technology upgrading. The latter includes modernization of existing plants and building new plants with the latest in automated technology. At each division, key technical personnel are working to become and are gaining recognition as Certified Quality Engineers.

AS is implementing a failure prevention plan or failure mode and effect analysis (FMEA) to prevent the production of defective parts, to identify defective parts, and to eliminate defective parts due to special causes. FMEA will be performed for each family of parts and/or production line to identify potential failure modes. Each potential failure mode is analyzed to determine the necessary controls and corrective actions. In other words, the process will be foolproofed against special causes by identifying conditions which could cause defective products.

AS's mechanism for implementing and monitoring its quality programs company-wide is its Quality Council (QC), a group of key design, production, engineering, and quality personnel from each company division.

The ASQC's focus is directed at four main areas: supplier quality, internal division audits, education and training, and self certification. The supplier quality program focuses on products and processes, with an ultimate goal of building long-term relationships with suppliers. A key strategy of AS's purchasing groups is to develop the quality systems of these long-term suppliers. A team was selected to define who should survey suppliers, how to consistently rate suppliers, when they should be surveyed, and how to communicate the results.

Supplier development guidelines were established which makes each division responsible for surveying all suppliers from which they are the largest consumer. All surveys are scored according to the guidelines, and results are published monthly to each division's purchasing manager. Results of a recent survey showed that 69% of the suppliers surveyed meet or exceed minimum standards for an acceptable quality system. Suppliers that do not meet the standards have shown willingness to make improvements.

Cross-divisional internal audits on quality systems are also conducted. This provides a mechanism for exchanging quality concepts between divisions and taking the strong points of each divisions's quality system and making them a part of other division's systems. All internal audits are conducted by teams using the same rating system that AS uses with its suppliers on at least an annual basis. Each division, at the conclusion of the audit, is provided a list of areas it should address to upgrade its quality efforts and these become excellent projects for improvements.

The ASQC is also developing a self certification program for AS products. Many AS customers currently have a program that allows each product or a supplying location to become certified. Once this designation is bestowed upon a supplier, their products no longer require incoming inspection but are shipped straight to the assembly line. This allows just-in-time production and delivery and reduces cost. Such a self certification program allows AS to present evidence that its products are worthy of use without inspection to customers who do not currently have a certification program of their own. This turns quality into an important marketing strategy.

Words and Expressions

productivity [prɔdʌk'tiviti] n. 生产力,生产率
manufacturing [mænju'fæktʃəriŋ] a. 制造的,制造业的;n. 制造,生产
Society of Manufacturing Engineers (SME) 美国制造工程师学会
foreign-built machine tool 外国制造的机床
trouble-free 无故障的,可靠的
service of equipment 设备的技术维护
offshore ['ɔ:fʃɔ:] a. 海外的,国外的,在国外建立的
overview ['əuvəvju:] n. 总的看法,概观,综述
term v. 称为,把…叫做
installation and startup 安装和启动
optical goods 光学产品
respondent [ri'spɔndənt] a. 回答的;n. 回答者
catch-up 迎头赶上
underway ['ʌndə'wei] a. 起步的,进行中的
in an effort to 企图,努力想
predominant [pri'dɔminənt] a. 占主导地位的,(在数量上)占优势的

valve [vælv] *n.* 阀
redouble [ri'dʌbl] *n.* 再加倍,加强
win back 恢复,夺回
quality program 质量保证计划,质量规划
upgrading ['ʌpgreidiŋ] *n.* 提高质量,改造,改进,升级
certified ['sə:tifaid] *a.* 持有证明书的,有书面证明的,经过检定的
FMEA = failure mode and effect analysis 故障模式和效应分析
foolproof ['fu:lpru:f] *a.* 十分简单,不会出错的
audit ['ɔ:dit] *n.* 审计,查账,审核
ASQC = Associated Spring Quality Council
quality system 质量体系,质量系统
designation [dezig'neiʃən] *n.* 牌号,命名,规定,目标
bestow [bi'stəu] *v.* 给予,赠予
just in time production 准时生产(即只在需要的时候,按需要的量生产所需的产品)

28. Quality in the Modern Business Environment

It is essential that products meet the requirements of those who use them. Therefore, we define quality as *fitness for use*. The term *consumer* applies to many different types of users. A purchaser of a product that is used as a raw material in its manufacturing operations is a consumer, and to the manufacturer fitness for use implies the ability to process this raw material with low cost and minimal scrap or rework. A retailer purchases finished goods with the expectation that they are properly packaged, labeled, and arranged for easy storage, handling, and display. You and I may purchase automobiles that we expect to be free of initial manufacturing defects or *nonconformities*, and that should provide reliable and economical transportation over time.

There are two general aspects of quality: quality of design and quality of conformance. All goods and services are produced in various grades or levels of quality. These variations in grades or levels of quality are intentional, and, consequently, the appropriate technical term is *quality of design*. For example, all automobiles have as their basic objective providing safe transportation for the consumer. However, automobiles differ with respect to size, appointments, appearance, and performance. These differences are the result of intentional design differences between the types of automobiles. These design differences include the types of materials used in construction, tolerances in manufacturing, reliability obtained through engineering development of engines and drive trains, and other accessories or equipment.

The *quality of conformance* is how well the product conforms to the specifications and tolerances required by the design. Quality of conformance is influenced by a number of factors, including the choice of manufacturing processes, the training and supervision of the work force, the type of quality-assurance system (process controls, tests, inspection activities, etc.) used, the extent to which these quality-assurance procedures are followed, and the motivation of the work force to achieve quality.

There is considerable confusion in our society about quality. The term is often used without making clear whether we are speaking about quality of design or quality of conformance. To achieve quality of design requires conscious decisions during the product or process design stage to ensure that certain functional requirements will be satisfactorily met. For example, the designer of an office copier machine may design a circuit component with a redundant element, because he knows that this will enhance the reliability of the product in the field and will increase the mean time between failures. This in turn will result in fewer service calls to keep the copier running, and the consumer will be far more satisfied with the performance of the product. Designing quality into the product in this fashion often results in a higher product cost. However, such cost increases are actually *prevention costs*, as they are intended to prevent quality problems at later stages in the life cycle of the product.

Most organizations find it difficult (and expensive) to provide the customer with products that have flawless quality characteristics. A major reason for this difficulty is *variability*. There is a certain amount of variability in every product; consequently, no two products are ever identical. For example, the thickness of the blades on a jet turbine engine impeller are not identical even on the same impeller. Blade thickness will also differ between impellers. If this variation in blade thickness is small, then it may have no impact on the customer. However, if the variation is large, then the customer may perceive the unit to be undesirable and unacceptable. Sources of this variability include differences in materials, differences in the performance and operation of the manufacturing equipment, and differences in the way the operators perform their tasks. Therefore, we may define *quality improvement* as the reduction of variability in processes and products. Since variation can only be described in statistical terms, *statistical methods* are of considerable use in quality-improvement efforts.

Quality is becoming the basic consumer decision factor in many products and services. This phenomenon is widespread, regardless of whether the consumer is an individual, an industrial corporation, or a retail store. Consequently, quality is a key factor leading to business success, growth, and enhanced competitive position. There is a substantial return on investment from an effective quality-improvement program that provides increased profitability to firms that effectively employ quality as a business strategy. Consumers feel that the products of certain companies are substantially better in quality than those of their competition, and make purchasing decisions accordingly. Effective quality-improvement programs can result in increased market penetration, higher productivity, and lower overall costs of manufacturing and service. Consequently, firms with such programs can enjoy significant competitive advantages.

All business organizations use financial controls. These financial controls involve a comparison of actual and budgeted costs, along with an associated analysis and action on the differences or *variances* between actual and budget. It is customary to apply these financial controls on a department or functional level. For many years, there was no direct effort to measure or account for the costs of the quality function. However, starting in the 1950s, many organizations began to formally evaluate the cost associated with quality. There are several reasons why the cost of quality should be explicitly considered in an organization. These include:

1. The increase in the cost of quality because of the increase in the complexity of manufactured products associated with advances in technology.

2. Increasing awareness of life cycle costs, including maintenance, labor, spare parts, and the cost of field failures.

3. The need for quality engineers and managers to effectively communicate the cost of quality in the language of general management—namely, money.

As a result, quality costs have emerged as a financial control tool for management and as an aid in identifying opportunities for reducing quality costs.

Words and Expressions

purchaser [ˈpəːtʃəsə] n. 购买人,买方,买主

nonconformity [ˈnɔnkənˈfɔːmiti] n. 不适合,不一致

quality of conformance 符合质量标准的程度,符合质量(用设计质量和制造质量,或预期质量和实际质量的差值来评价产品制造质量的水平)

have as 把…当作

appointment [əˈpɔintmənt] n. 指定,家具,车身内部装饰

intentional [inˈtenʃənl] a. 故意的,有意的

driven train 驱动轮系,动力传动系统

accessory [ækˈsesəri] a. 附属的,附带的,n. 附属品,附属装置

quality assurance 质量保证

motivation [məutiˈveiʃən] n. 刺激,激发,诱导,动机

redundant [riˈdʌndənt] a. 冗余的,过量的,重复的 n. 多余部分,备份

mean time between failures 平均故障间隔时间

prevention cost 预防费用

flawless [ˈflɔːlis] a. 无瑕的,无缺点的,完美的

variability [vɛəriəˈbiliti] n. 易变性,变化性,变异度

jet turbine engine 喷气涡轮发动机

impeller [imˈpelə] n. 叶轮,涡轮,推进器

quality improvement 质量改进

statistical [stəˈtistikəl] a. 统计的,统计学的

market penetration 市场渗透,市场份额

it is customary to 通常,一般习惯于做

29. Design and Manufacturing Tolerances

Tolerance is one of the most important parameters in product and process design, and it plays a key role in design and manufacturing. Tolerance is defined as the maximum deviation from a nominal specification within which the part is acceptable for its intended purpose. A tolerance is usually expressed as lower and upper deviations from the nominal value.

Manufacturing involves applying a series of operations to components (parts, subassemblies, etc.). These operations are intended to ensure specific geometry on workpieces. Dimensions in engineering drawings specify ideal geometry: size, location, and shape. Because variations exist in both process and material, the manufacturing process creates a part that has an approximate geometry of the ideal. Tolerances are introduced to specify and control the variations. With the advent of assembly lines, it became critical to manufacture interchangeable parts. The use of replacement parts for maintenance operations also requires the interchangeability of parts. Tolerances are used to ensure that parts have this property.

Two types of tolerances are often used: design tolerances and manufacturing tolerances. Design tolerances are related to the operational requirements of a mechanical assembly or of a component; whereas manufacturing tolerances are mainly devised for a process plan for fabricating a part. Manufacturing tolerances must ensure the realization of design tolerances. For example, three manufacturing processes are used to make a $20^{+0.01}_{0}$ mm hole. The processes include drilling, boring, and grinding. Assume the following dimensions and tolerances must be maintained in these processes; $18.5^{+0.2}_{0}$, $19.5^{+0.05}_{0}$, and $20^{+0.01}_{0}$. Then $20^{+0.01}_{0}$ is called the design dimension and tolerance for the hole. The manufacturing (machining) dimensions and tolerances for the drilling, boring, and grinding processes are $18.5^{+0.2}_{0}$, $19.5^{+0.05}_{0}$, and $20^{+0.01}_{0}$, respectively.

There are two kinds of tolerancing exercises performed in product and process design: tolerance analysis and tolerance synthesis. Tolerance analysis involves identification of related tolerances in a design, and calculation of the stackup of these related tolerances. The process of tolerance accumulation is modeled, and the resultant tolerance is verified. If design requirements are not met, tolerance values are adjusted and the stackup recalculated. Tolerance synthesis, on the other hand, is the process of allocating tolerance values associated with design requirements in terms of functionality or assemblability among identified related design (or manufacturing) tolerances. It is a process of distributing tolerance values among a number of related dimensions.

A tolerancing problem can be solved based on either a worst-case approach or a statistical approach. In the worst-case approach, the extreme values of tolerances are considered. Complete interchangeability is ensured for all in-tolerance parts. Because the machining of parts is performed on different machine tools, or on the same machine tools but at different points in time, the dimensions on the parts are independent stochastic variables. Thus statistical tolerance calculation is valid. In the statistical approach, tolerance calculations are performed based on the fact that actual part dimensions are randomly distributed about their nominal values. While allowing larger tolerances to be used in design and manufacturing, statistical tolerancing leads to lower manufacturing costs. However, scraps will be produced as the result of the statistical method. The worst-case tolerance synthesis applies to single piece or small-lot production for critical applications, e. g. , aerospace aircraft; whereas the statistical method makes much more sense for mass production.

Both design and process engineers are concerned about the effects of tolerances. Designers usually specify tight tolerances to ensure the performance and functionality of the design. Process engineers prefer loose tolerances, which make parts easier and less expensive to produce. Therefore, tolerance specifications become a critical link between design and manufacturing. Good tolerance design ensures quality products at low cost.

For mechanical parts, the manufacturing error on a dimension primarily depends on the following factors: accuracy of machine tool; accuracy of fixture; accuracy of tool; setup error; deformation of the machining system (including machine tool, fixture, tool, and workpiece) under external forces; thermal deformation of the machining system; measurement error; impurity of material.

These errors can be grouped into two categories: deterministic and random. For example, the manufacturing error on the diameter of a reamed hole due to the inaccuracy of the reamer diameter is a deterministic error. The manufacturing error caused by the variation of material hardness is a random error. The total manufacturing error on the dimension of a part feature is the combined effect of the above-mentioned errors. These errors are mainly related to the inaccuracies of machine tool, fixture, tool, setup, gage, and material. In addition, the operator skill level is a factor that affects manufacturing error. In a manufacturing firm, there usually exist a number of alternate combinations of machine tools and auxiliary equipment for a manufacturing process. Each combination is associated with a manufacturing precision level and cost. A higher precision level (tighter tolerance) usually requires a higher manufacturing cost, due to the need for a more accurate machine tool, fixture, tool, setup and more skillful operators. The precision of the process varies, depending on the accuracy of resources (including equipment, material, and operator). Therefore, for a specific manufacturing process, there is a monotonic decreasing relationship between manufacturing cost and precision in a certain range.

Manufacturing tolerance allocation is to determine properly the manufacturing tolerances in intermediate manufacturing processes for a part fabrication. A mechanical part usually has a number of dimensions and thus the machining of the part is for the realization of those dimensions. In manufacturing practice, a dimension is usually obtained by performing several manufacturing processes. The number of processes depends on the complexity of geometry and the tolerance—surface roughness requirements on the dimension and its related surfaces. The relationships between machining dimensions involved in a part fabrication are often complex, especially when a large number of operations are used to make the part. Tolerancing charting has been used to establish the relationships among the machining dimensions. It provides the process engineer with an effective tool for machining dimension and tolerance analysis.

Words and Expressions

nominal ['nɔminl] a. 标定的,额定的;n. 标称,额定
deviation [di:vi'eiʃən] n. 偏差,偏移,差异,误差
lower and upper deviation 下偏差和上偏差
operation 工序
interchangeable [intə'tʃeindʒbl] a. 可互换的,可拆卸的,通用的
replacement 替换,替换物
devise [di'vais] v. 设计,发明
process plan 工艺规程,生产工艺设计
boring ['bɔ:riŋ] n. 镗孔,镗削加工
synthesis ['sinθisis] n. 合成,综合,拼合
stackup ['stækʌp] n. 堆积,层叠

accumulation [əkju:mju'leiʃən] n. 积累,累加,累积
resultant [ri'zʌltənt] a. 合成的,作为结果发生的;n. 生成物
distribute [dis'tribju:t] v. 分配
worst-case approach 极值法
statistical approach 概率法
extreme [ik'stri:m] n. ;a. 极端(的),极度(的);n. 极端的事物
complete interchangeability 完全互换
independent variable 自变量
stochastic [stə'kæstik] a. 随机的,不确定的,偶然的
randomly ['rændəmli] ad. 任意地,偶然地
scrap [skræp] n. 废品
small-lot production 小批量生产
functionality [fʌŋkʃən'æliti] a. 功能性
deterministic [ditə:mi'nistik] a. 确定性的
random ['rændəm] a. 随机的
ream [ri:m] n. ;v. 铰孔,铰削
reamer ['ri:mə] n. 铰刀
monotonic [mɔnə'tɔnik] a. 单调的
tolerancing charting 公差的图表计算法

30. Product Reliability Requirements

Is reliability a saleable commodity? Yes and no. It all depends on the type of industry you are in. In the aerospace industry one can give an unqualified yes. Considering the technical complexity of their missions, the reliability achieved by the Apollo space missions is remarkable. One hundred per cent safe returns of the Apollo astronauts must be the reliability success story of the second half of the twentieth century. Did someone say why? Simply because reliability was written into the specification along with other performance parameters at the conceptual stage of the project. Tenders were therefore on the basis of achieving the standard of reliability written into the specification.

In the global market economy, the incentive to strive for a high level of reliability is simply one of profit. The individual firm or organization must assess the level of reliability necessary to enable the required sales targets to be achieved, along with the other engineering parameters of performance and operating characteristics. The production costs that must be maintained to enable the product to be put on the market at a competitive selling price will usually dictate the type of construction to be employed. Where this is novel, then the engineering development costs must allow for the spending associated with achieving the required level of reliability. Launching a new product which has an unsatisfactory reliability can make a complete nonsense of the best-laid marketing plans. The development work can, in many cases, be used as part of the marketing strategy, and can form an essential part of the launch programme. It is also good insurance if the initial reliability is shown in the specification, for the customer is usually much less irate if he believes that the trouble has struck in spite of a sensible and visible programme of work designed to check out the reliability of the product.

Examples of the combined development and marketing strategy come from the automotive industry, the aero-engine industry, and the fan industry. When a new model of car is announced it produces two opposite reactions in the potential customer's mind:

(1) It is new (and therefore "better" than the previous one).

(2) It is new (and therefore unproven and potentially full of teething troubles that will give unreliable operation).

The strategy is to make the customer think it is new (and therefore better than the previous one and therefore desirable) and it has been well proven by test (and therefore should have few hidden faults which will require frequent visits to the service station). To do this it is vital to include in the publicity the more readily understood tests to prove the reliability of the new car, such as "this car has been driven for 100 000 miles before we let the public know of its existence," or "two hundred selected motorists have been given this car to test to make sure that our new car will stand up to the rigors of everyday business motoring." Both of these examples have been used in the past to try and involve the motoring public in the extensive development work that has been carried out on their behalf to arrive at a desirable and reliable product.

The strategy in the aero-engine industry is similar in that there must be a convincing reply to the airline or armed-forces customer who says that he can remember the teething troubles he had with the last new engine, and what has been done to make sure that he does not have to go through a similar period all over again? The simplest way to put over the development programme is to show that with the previous new engine, 10 000 hours of engine running were carried out on the test bed before entering airline or military service, but with the new engine, 14 000 hours of running will be carried out on the test stand before service operation begins. In this sort of situation it is absolutely essential to use the very best market intelligence about what the customer really wants, and is really prepared to pay for.

In the fan (see Fig. 30.1) industry the customer is concerned with price, performance, and reliability, and usually in that order. There is therefore likely to be less money in the kitty for elaborate testing, and every test must be meaningful to the engineer and convincing to the customer that he can install the product and forget it. Field testing is therefore very important for it is relatively cheap, but of course takes a fair amount of elapsed time. The customer here is particularly convinced by seeing installations, or details of installations, where the new product is running satisfactorily. Damp, steamy, or dirty installations are particularly desirable to introduce an element of "overstress" testing into the field service trials.

Figure 30.1 Fans

If the fan is to be installed in an agricultural situation, then it is important not to field test the fans with the most careful user that can be found, but rather with one who cleans down in a more haphazard fashion, hosing down the equipment when he should not. The hosing down may well show that a sealing arrangement needs to be incorporated on the shaft to stop water entering the motor and ruining the reliability.

Words and Expressions

commodity [kə'mɔditi] n. 物品,商品,货物
reliability [ri,laiə'biliti] n. 可靠性,安全性,准确性
unqualified ['ʌn'kwɔlifaid] a. 不合格的,无条件的,绝对的
specification [,spesifi'keiʃən] n. 尺寸规格,技术要求
tender ['tendə] n. 招标,投标
irate [ai'reit] a. 发怒的,愤怒的
teethe [ti:ð] v. 出牙
teethe troubles 事情开始时的暂时困难,初期困难
publicity [pʌb'lisiti] n. 宣传材料,广告
motorist ['məutərist] n. 汽车驾驶员,开汽车的人

motoring ['məutəriŋ] n. 驾驶汽车; a. 汽车的
fan 通风机(是依靠输入的机械能,提高气体压力并排送气体的机械)
kitty ['kiti] n. 凑集的一笔钱,共同的资金
elapse [i'læps] v. ; n. (时间)过去,消逝
overstress [əuvə'stres] n. ; v. 过载,超载,过度应力
haphazard ['hæp'hæzəd] n. 偶然性,任意性; a. 杂乱的,任意的
hose down 用水龙带冲洗,用软管洗涤

31. Product Reliability

Reliability is sometimes simply defined as the probability of a product or process performing its intended or specified function. Inherent in this definition is the implication of performing under certain stated conditions or environments as well as performing for a specified length of time.

Quality has a little different definition, and it is sometimes defined as "the totality of features which determine a product's acceptability." Some may choose to define this as conformance to specifications, fitness for use, or meeting requirements the first time. Reliability, as a concept, goes beyond the basic ideas in quality because it adds the concept of time to the basic definition. That is, reliability may also be viewed as meeting requirements, the first time and every time. Products may have quality yet not have reliability.

Many companies have recognized that quality by itself is not sufficient in a competitive world. Products must be built and designed to last a long time in their intended function. This additional requirement goes beyond basic quality and adds a new dimension to marketing for the company. The most successful companies, in terms of product acceptance, have recognized the importance of total life cycle cost and customer satisfaction. A good understanding of reliability is necessary to achieve both with minimum cost in terms of time and resources.

There are three elements that provide the basis for a reliable product—design, manufacturing, and component parts. Design is that series of operations involved in taking a product from a conceptual stage to a form that meets both company goals and customer expectations. In addition, this step should include some demonstration that the goals have indeed been met. This is often referred to as the design validation step. Typically, a limited number of key features of the product are demonstrated to meet the goals over the expected life of the product. Reasons for success as well as failure can then be identified early in the product's history and corrected or improved before cost and schedule become big constraints.

Reliability techniques are needed to determine the most appropriate and effective tests as well as reduce the test time, while preserving test conditions. This is known as accelerated testing. Teamwork becomes important here. The object is to finish the design process successfully in as short a time as possible. Bringing other groups into the design validation process will help minimize the number of tests needed later and will reduce duplication.

The next step, manufacturing, involves turning the design into reality without affecting it adversely. Consistency, via such techniques as statistical process control (SPC), is the key to creating products that can be tested and demonstrated as being reliable. More than one reliable design has been affected by a process that damaged or weakened one or more of the components or subassemblies. This often occurs without anyone's knowledge, until a later test or a user of the product identifies the situation.

The third important element involved in producing a reliable design includes the parts and subassemblies that go into the final product. High-quality and consistent parts are needed to preserve the design and be compatible with the established manufacturing processes. This means selecting the key parts suppliers ahead of time, qualifying each of the part types, then not changing during the life of the product without some requalification. Supplier-initiated changes in purchased items should be discouraged. In order to prevent such a situation, a clear note—" No change of process without prior notification"—is needed on the procurement documents.

The reliability function includes estimating conformance to an anticipated life, as well as warranty and failure analysis information. This role establishes the reliability group as a critical interface between many groups.

When developing a test plan or conducting failure analysis, teamwork becomes important. Typically, development of the original plan to test the final product involves marketing, design, manufacturing, and product management. A meeting of representatives from all of these functions to discuss a proposed test outline will almost always produce a test plan that meets the design intent and identified customer needs.

One valuable model used by reliability engineers is the so-called bathtub curve(Fig. 31. 1), which is named for its particular shape. This curve often describes the reliability of a complex system. The descending slope near the start represents the early history of an item and indicates a decreasing failure rate. It indicates that the first few hours or days of a product's life are the worst, followed by a period of improvement. This is sometimes referred to as the infant mortality period. For electronic systems, it could last 5000 hours, or more.

Over time, the failure rate decreases to a low point where it remains essentially constant for a long period. This portion of the curve

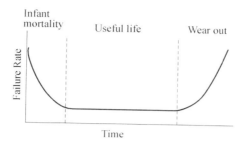

Figure 31.1 Bathtub curve

is called the useful life. It is the constant failure rate period of a product's life. The last portion of the curve depicts an increasing failure rate and is attributed to wear out. Not surprisingly, the highest number of failures occur during this period. Properly used, the bathtub curve can also shed valuable light on what to expect when a product or system is ultimately put to the test in the field.

Reliability and quality go hand in hand in a variety of ways. Both are vital in maintaining a competitive position in the world marketplace. It is important to understand the basic concepts and methods for improving product reliability and the ground rules for applying them at all stages of product design and manufacturing.

Words and Expressions

totality [təu'tæliti] n. 全体,总数,总额,完全
attribute ['ætribju:t] n. 属性,特征,标志,象征
validation [væli'deiʃən] n. 使生效,合法化,批准
adversely ['ædvə:sli] ad. 有害地,不利地,相反地
via [vaiə] prep. 经过,取道
statistical process control(SPC) 统计过程控制
without anyone's knowledge 没有任何人知道
identify [ai'dentifai] v. 识别,确定,发现

ahead of 在…前头,超前于,比…提前
notification [nəutifiˈkeiʃən] n. 通知,告示,布告
warranty [ˈwɔrənti] n. 保证(书),担保(书)
mortality [mɔːˈtæliti] n. 致命性,失败率,死亡率
infant mortality 早期失效率
depict [diˈpikt] v. 描述,描写
bathtub curve 浴盆曲线(在产品整个使用寿命其间,典型的失效率变化曲线,形似浴盆)
useful life period 使用期
wear out 耗损,消耗,耗尽,用坏,用完
wear out period 耗损失效期,磨耗期
go hand in hand 紧密相关,相随相生,并驾齐驱

32. Total Quality Management

Many manufacturers in Europe and the United States experienced a loss in market share in the mid-1970s; even producers who were able to maintain their market share felt the pressure of foreign competition. At the same time, manufacturing technology started to mature with the introduction of sophisticated and reliable hardware such as robots, computer numerical machine tools, programmable logic controllers, material-handling systems, and coordinate-measuring machines. Computers became smaller and less expensive. In addition, the number of off-the-shelf software solutions for manufacturing and production control increased dramatically.

The lesson learned by manufacturing in the 1970s is that technology alone cannot improve business performance, and a new emphasis on quality and development of the workforce was required. The move to quality has two major elements: a new philosophy on how the business must function and the use of quality tools to achieve the desired results. This broader view of quality is included in a process called total quality management (TQM). TQM has two components: principles and tools. The principles allow an organization to overcame the traditional barriers that prevented the management group from utilizing the potential, skills, and knowledge of every employee. The tools permit quantitative and qualitative measurement of the system to determine how well the process is meeting organizational goals.

The definition of TQM has three dominant themes: (1) participative management for everyone in the enterprise, (2) implementation of a successful continuous improvement process, and (3) efficient use of multifunctional teams. The six important principles that must be considered in a TQM implementation include (1) a focus on customer needs and satisfaction; (2) a focus on the process used to produce a product as opposed to a result or product focus; (3) a focus on the prevention of problems in areas such as quality, production, machine operation, and engineering design changes; (4) a focus on using the brain power of every employee in the organization; (5) a focus on basing decisions on facts about manufacturing and design without blame when problems occur; and (6) a focus on developing an enterprise where communication channels are always open so that product data and process information flow freely among all levels and all employees.

The goal of producing with near zero defects is reached through the adoption of a six-sigma (6σ) design and production process. The products from most manufacturing processes have a normal distribution, which means that 99.74 percent of all production falls within plus or minus three standard deviations (± 3 sigma) of the average center value. The relationship between the normal curve and the product and process design determines how the process variation will affect the final product. The relationship between process variation and design tolerance is shown in Figure 32.1. The ±3 sigma range for the process variation is called the process capability, and defects occur when this process capability is wider than the design tolerance (T) or specification width marked by the lower specification limit (LSL) and upper specification limit (USL). Anything that causes the process capability to exceed either specification limit (LSL or USL) causes defects in the product.

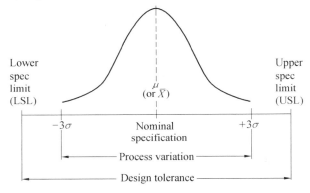

μ = mean or average value
σ = standard deviation

Figure 32.1 Process variation and design tolerance

The process capability index (Cp), a measure of the inherent capability of the process to produce parts that meet the design requirement, is found by dividing the design tolerance by the process capability ($Cp = T/6\sigma$). The larger the process capability index, the more likely that the parts will satisfy the production requirement. One method of ensuring defect-free parts would be to make the design tolerance as wide as possible and the process capability narrow.

If machines produce parts with the standard variation found in a normal distribution, most of the parts will fall between ±3 standard deviations of the average value. If the allowable specification width for critical part dimensions is ±6 standard deviations, even with normal process variation fewer than four defective parts are produced in every batch of 1 million. This process leads to near-zero-defect production.

Management of the automated workforce is changing as a result of two key internal factors: the relationship between the operator and the machine, and the highly integrated nature of automated production systems. External factors affecting workforce management are the expanding global marketplace and evolution of the workforce. The most significant event resulting from evolution of the workforce is the move to self-directed work teams. The move to these teams is a result of two factors: first, for the organization to remain competitive, global competition requires problem solving from everyone in the enterprise; and second, the workforce has evolved to a state where employees are ready to take the responsibility to control the production processes assigned to them by management. After self-directed work teams are implemented, management becomes the "what" team to decide what and how much to make; the self-directed work teams deal with the "how" issues, where they focus on how to produce the product most efficiently.

Words and Expressions

material handling 物料搬运
coordinate measuring machine 坐标测量机
off-the-shelf 现成的,通用的
business performance 企业经营业绩
total quality management 全面质量管理
quantitative ['kwɔntitətiv] *a.* 数量的,定量的
qualitative ['kwɔlitətiv] *a.* 定性的,性质上的

theme [θi:m] n. 主题
as opposed to 与……相反,与……相对比
design change 设计变更
brain power 脑力,智能
communication channel 信息通道,信息交流通道
zero defect 零缺陷
normal distribution 正态分布
normal curve 正态曲线
process variation 加工偏差,加工变化范围
process capability 工序能力,设备加工能力
specification width 规定的尺寸范围,技术要求范围
lower specification limit 尺寸下限,技术要求下限
process capability index 工序能力指数
self-directed 自我指导的,自主的
production process 生产过程

33. Employees Take Charge of Their Jobs

How much would you spend for a system that increases your plant's productivity by 30%, cuts labor cost in half, and reduces scrap and rework to practically zero? Sound too good to be true? Want proof? Read on.

Back in 1985, Lord Aerospace in Dayton, OH, invested in a system that by 1990:

- improved productivity by 30%,
- reduced scrap and rework cost by 85%,
- improved cycle time for one part from 75 days to seven,
- reduced a 50% reject rate on parts shipped to near zero.

So what is this system, and how much does it cost? Lord calls it "self-directed work teams," and its costs are measured in terms of communication, dedication, involvement, and trust rather than in dollars.

Lord Corp's Dayton plant is a part of the Lord Aerospace Products Division. The larger portion of the aerospace operation is located in Erie, PA. Dayton employs approximately 70 employees distributed across six self-directed process teams and an additional three support teams. There are also some teams involved in activities such as newsletters, safety, and security. The plant runs two 10-hour shifts, manufacturing vibration-isolator mountings for aircraft companies such as Boeing, Lear Jet, and Piper. It is also Lord Aerospace's most productive plant.

In the beginning

Back in 1985, Lord was close to shutting down the Dayton facility. Problems such as a 50% reject rate on parts received from Dayton and a high employee turnover rate were becoming too much to tolerate. "One year we had 57 employees and we turned over 45 people. The discharges, job eliminations, layoffs, and quits all made a real poor environment for everyone," says Mike Rogers, employee relations manager at Lord-Dayton.

So how does a company on the brink of closing its doors back in 1985 not only turn a plant around but transform it into one of the company's most productive plants? "We began by going to the workers, the experts," said Mr Rogers. "We asked them what it would take to turn this place around."

Some of their answers surprised management, says Dave Lichtinger, plant manager. "Basically, what workers told us was that they were over managed. While the company had been expounding the virtues of empowerment and employee involvement, the reality was that employees were hampered by management."

Management delegated responsibility, but not authority. "There were too many tiers of management," said Mike Smedley, one of the team leaders at the plant. "There were so many layers of management and so many bosses that at one time it took six signatures to get a tool from the crib. Employees lacked motivation and merely did what they were told."

Soon after, conditions began to change when work teams were formed to improve working conditions and solve problems.

"We were hearing some pretty stark things," says Mr Rogers, "how heavy-handed we were, how we really didn't give workers any freedom, and how there wasn't any trust."

Out of those meetings came a five-year plan that would eventually culminate in an all-equal operation in which the workers run their operation as if they were running a business.

"When we submitted that plan, there were a lot of skeptical people," continues Mr Rogers. "There were those who thought we should have been able to turn it around in a year or two. But we felt there was a tremendous environment change or culture change needed and it was going to take a while. In fact, it was just recently (1990) that we went all-salary. To build trust it really took a lot of time and a lot of working together instead of the old 'us and them' attitude."

The transition

Over the first five years of the plan, many changes took place, all initiated and directed by the workers. The first step, in 1986, was to develop worker involvement teams and a no-layoff pledge from management. "We made a pledge that no one would be laid off as the result of any improvements in the processes," says Mr Rogers. "Barring some horrible economic down-turn in our market place, we would not have any layoffs."

At first, involvement teams worked on reducing scrap and setup times. By 1987, the first workcell team was formed. "The team consisted of three operators, a CNC programmer, a facility engineer, and a maintenance technician that designed the cells themselves. They measured the floor, laid out the cell on paper, and we brought in riggers to move equipment around," said Mr Lichtinger. The resulting cell consisted of two Cincinnati Milacron T10 machining centers; a Mori Seiki SL7B turning center; a six-spindle drill press; and a deburring station to produce a complicated, tight tolerance helicopter rotor head component from forging to finished part. Until then, the part was the plant's worst quality problem.

The first workcell was a glowing success. Team members reduced scrap from over $300,000 a year to practically nothing. Leadtime was cut from 160 days to 32 on the cell's main product.

Because the pilot cell was so successful, the manufacturing cell concept gained a lot of credibility and attention at Lord-Dayton. "We used what we learned to quickly form other cells, with just as much success," says Mr Lichtinger. "In 1989, a cell was formed using five Toyoda FH55 horizontal machining centers, a deburring station, and a stamping press to produce a series of engine-mount components."

By mid-1989, five self-directed work teams were established, eliminating four supervisors from the shop floor. Today, the company has six self-directed production teams consisting of from four to eight people, and three support teams. The plant consists of four machining cells, an elastomer processing/bonding cell, and an assembly/packing/shipping cell.

Much emphasis has been placed on integrating previously indirect functions into the cells, thereby eliminating support personnel, and increasing the autonomy of each cell.

Building the teams

Does it take a special kind of person to work in the Lord environment? Are there unique skills or personality traits required to fit the mold? "We look for somebody who has X number of years experience and/or a graduate of one of the technical machine schools in our area such as our county's Vocational School or Community College." says Mr Rogers. "Then we bring in at least five candidates to be interviewed by a selection team. If the opening is in a manufacturing cell, there will be two cell team members, a quality person, a tooling person, an engineer, myself, and maybe someone from materials. One by one, we interview the five people and eventually we get together and we go around the table and rank them. We're pretty much able to draw a consensus on the number-one candidate. Everyone gets an equal vote. When the candidate is brought on board, the team leader assigns a team member as a trainer."

Last year Lord invested about 3000 hours of training in its workforce. The company is big on cross-training so that every member of the team is able to perform all jobs the team is responsible for. A new member may require two or three years of training depending on their experience level coming in.

The company looks for team players. People who are loners cannot function for long in the Lord system "You can't look out just for yourself," says Kerry Lynch, machinist. "You have to look at the whole team concept to make sure that your team is going smoothly and working as a unit together."

Personnel problems are also handled at the team level. Management enters the picture only if a team requests it. For example, if an individual is unable to get along with other team members, or two members of an individual team have a problem with each other, the team would ask Mr Rogers to become involved. "I might do some formal counseling based on team input. The individual being counseled realizes it's the team members' input as well as my observation, and if it goes to several counseling sessions, it could lead to termination," says Mr Rogers.

The company also maintains a RAP (Resource Assistance Program) line to assist troubled employees in finding help, either supplying or referring them to needed counseling. That gives any member an opportunity to go outside for help. Whether the problem is a child on drugs, alcohol problems, or marital difficulties, they can call the RAP line, seven days a week, 24 hours a day.

Despite having to overcome deplorable working conditions, poor moral, and mismanagement, the people at Lord-Dayton have managed to create an American success story. Work cells, new equipment, and a recent $8 million expansion have positioned the company well in today's marketplace as a world-class supplier to the aerospace industry. At the root of their success, however, are still the people. As Mr Rogers summed up: "We can have the nicest building in the world and state-of-the-art technology, but if you don't have the people, you are not going to be successful. That's what it really takes, giving people the power to be in charge of their jobs."

Words and Expressions

take charge of 负责
rework [ˈriːwəːk] n. 重新加工,返工,修改
self-directed 自我指导的,自主的
self-directed work team 自我管理团队
newsletters 通讯稿,业务通讯,简讯
vibration-isolator 振动隔离器

facility [fə'siliti] n. 工厂,设备
turnover ['tə:nəuvə] n. 人员调整,人员更新
discharge 解雇,开除
layoff (暂时)解雇,下岗
quit 辞职,辞退
on the brink of 在…边缘,濒于
turn...around 改变,转变
expound [ik'spaund] v. 详细说明,解释
hamper ['hæmpə] v. 妨碍,阻碍,阻止
delegate ['deligit] n. 代表;v. 委派,授权
tire [tiə] n. 层;v. 分层布置
crib [krib] n. 木笼
stark [stɑ:k] a. 僵硬的,苛刻的
heavy-handed a. 严厉的,不明智的
out of 由…产生出
culminate ['kʌlmineit] v. 达到顶点,结束
culminate in 以…而终结,以…而达到顶峰
skeptical ['skeptikəl] a. 怀疑的,怀疑论的
barring ['bɑ:riŋ] prep. 除…以外,不包括…
down turn 向下,下降趋势,衰落
rigger ['ri:gə] n. 装配工
turning center 车削中心
drill press 钻床
glowing ['gləuiŋ] a. 生气勃勃的
pilot 示范的,实验的
credibility [kredi'biliti] n. 可信性
stamping press 冲床,冲压机
elastomer [i'læstəmə] n. 弹性体,合成橡胶
bonding ['bɔndiŋ] n. 连(焊,胶,粘)接,粘结剂
packing and shipping 包装和运输

personality trait 个性品质，人的性格
vocational school 职业学校
community college 社区学院
opening 空缺，机会
consensus [kən'sensəs] n. 一致同意，多数人的意见
workforce 劳动力，职工总数
(be) big on 对…偏爱
cross-training 交叉培训，交替培训
team player 能共同努力和相互合作的人
counseling ['kaunsəliŋ] n. (对个人，社会以及心理等问题的)咨询服务
termination [tə:mi'neiʃən] n. 终止，结束
on drugs 吸毒
marital ['mæritl] a. 婚姻的
deplorable [di'plɔ:rəbl] a. 可悲的，悲惨的
world-class 世界级的，世界水平的
at the root of 是…的根本

34. Accelerating Development with Acquired Technology

For both small and large companies, it may be more cost effective to acquire technology, either through licensing or through joint ventures. This approach can lead to several benefits, including: a reduction in both costs and risks, an expansion of markets, an acceleration of sales and thus an increase in profits, and others. Perhaps the most important advantage of such a beneficial alliance and long-term relationship with the partner is "getting there first." In other words, one has the ability to preempt the competition from the market.

Before one acquires technology, one must find it and there are many ways to accomplish this. A patent search is perhaps the most obvious, while hiring a consultant is another alternative. One can rely on personal contacts and trade fairs as well. Both federal laboratories and universities generate new technology; they usually have a technology transfer office that handles licensing and joint ventures. However, each university may require a different approach in finding out what technology is available.

There are several rules of thumb that must be followed for successful acquisition of new technology. One must understand the implication of relevant trends and how these trends impact both the technology and its markets. Obviously, a company must also be receptive to new technology and avoid the "not invented here" syndrome that can be prevalent with top management. Once this attitude has been eliminated, an audit of resources (both funding and personnel) must be completed to determine what technology is needed. Of course top management must be totally committed and supportive throughout the whole acquisition process; whoever is in charge of technology acquisition (whether it be a committee or division) must have a direct line of communication to senior management. A company must be able to act quickly to capitalize on new developments and must have a complete understanding of its previous failures and successes.

A company must also realize that the market will have to be developed and the need for the technology verified. One must work closely with customers. A thorough understanding of the market is needed—both competitors and potential customers. The company must also make sure it has the compatibility and the manufacturing capacity to accommodate the technology. A sales plan is another critical step to the success of the product. One must determine how to introduce the product to the market and how to grow this market. However, a company must build inventory before the product is introduced because if the demand cannot be met, the competitors will win. One must also take into consideration that scale up of production will not be easy and, therefore, product quality must be monitored before introduction and after scale up. Perhaps the most important ingredient of a successful acquisition is the sharing of profits with the other partner, thereby creating a "win/win" situation for everyone involved.

One of the important criteria for success is compatibility. A joint venture should be considered in the same way as a marriage—determining if the relationship is worthwhile at the beginning. If good rapport can be achieved, then the partnership should be a success as long as there are comparable contributions in the areas of funding, marketing, technology, and manufacturing between the parties involved. Otherwise, one company will dominate which generally leads to failure, though there are certain exceptions where one company should dominate.

Total commitment from top management is again required to make the joint venture work. Personal relationships must be developed at all levels between the companies involved. It is also recommended for a large company to have a centralized marketing division to look at all businesses the company serves. This, in addition to having a division of technology assessment headed by a director, should facilitate the acquisition and marketing of new technology.

Words and Expressions

acquired [ə'kwaiəd] *a.* 已得到的,已获得的
cost effective 有成本效益的,划算
license ['laisəns] *n.* ;*v.* 许可(证),特许,执照,批准,认可
joint venture 合资,合资企业

preempt [pri'empt] v. 优先购买,先取,先占
patent ['pætənt] n. 专利
trade fair 商品交易会
rule of thumb 单凭经验来做的方法,经验法则
implication [impli'keiʃən] n. 关系,隐含,意义,本质,实质
relevant ['relivənt] a. 有关的,相应的
receptive [ri'septiv] a. 接受的,有接受力的,易于接受的,容纳的
syndrome ['sindrəum] n. 综合症,并发症
prevalent ['prevələnt] a. 流行的,盛行的,普遍的
top management 最高管理部门,高层管理人员
compatibility [kəmpætə'biliti] n. 相容性,兼容性,适合性,一致性
accommodate [ə'kɔmədeit] v. 使适应,接纳
inventory ['invəntəri] n. 清单,商品目录,存货
scale up (把)…按比例增加,递加,增大比例
worthwhile ['wə:ð'hwail] a. 值得做的
rapport [ræ'pɔ:] n. 关系,联系
party (诉讼,协约,会议等)一方,当事人
dominate ['dɔmineit] v. 支配,占优势
centralized 集中的
marketing division 销售部门
assessment [ə'sesmənt] n. 评估,评价,鉴定
facilitate [fə'siliteit] v. 推动,促进,帮助

4 MANUFACTURING AND AUTOMATION

35. Careers in Manufacturing

Manufacturing gives jobs to people and makes the products consumers need. Manufacturing companies hire people with many types of skills and talents to help the companies compete well in the world of business. Manufacturing companies often divide themselves into several departments. A department is a small part of the company.

Some of the more common departments found in large manufacturing companies include management, engineering, production, marketing, finances, and human resources. People with different talents, skills, and educational training can find many different types of jobs and careers in each of these departments.

Management Department. For a manufacturing company to be successful, all of the departments must work together toward one goal. Management's job is to make sure all the departments work together. Workers in management, called managers, make company policies and then make sure the other departments follow these policies.

Engineering Department. The engineering department designs, produces, and tests the first model of a new product. These first models are called prototypes. In the engineering department, scientists conduct tests and experiments on new materials like plastics, ceramics, and metal alloys. Technicians also work in the engineering department. A technician is usually a skilled and experienced design or production worker. Only the most experienced and skilled technicians can work in the engineering department as drafters, designers, and machinists. Engineers are the problem-solvers when it comes to designing a new product or the system for making that product.

Many engineering departments have a research and development team. This team is made up of the best engineers, technicians, and production workers. Their job is to create and test new product ideas.

Production Department. The production department sets up and uses the tools and machines to make the product. A management worker, the supervisor, is normally in charge of this department. The supervisor tells the production workers what to do and then makes sure they do it. Supervisors also train new workers and encourage safe work habits. Frequently, supervisors begin as production workers.

Production jobs use skilled, semiskilled, and unskilled labor. Skilled workers design and make specialized tools, jigs, fixtures, and quality control devices used to manufacture the product. They also adjust and maintain equipment, fixtures, or quality control devices when they break down. Because their jobs require precision skills, these workers usually have years of manufacturing experience and training. Skilled workers also set up and operate a variety of machines, including computer-controlled devices such as lathes, mills, and even robots. Semiskilled workers run machines and use the special tools, jigs, and fixtures that are made and set up by skilled workers. Unskilled workers perform physically demanding, routine jobs. They include material handlers, helpers, and assemblers. Unskilled workers do not have the skills to run machinery.

Marketing Department. The marketing department conducts market research and advertises, packages, sells, and distributes the product. Market researchers do surveys to see what consumers want, if they like their product, and what price they will pay. They also compare their products to those of competitors.

Advertisers decide which media will be used to publicize the new product. Artists, photographers, and writers prepare advertising campaigns for print media like newspapers, magazines, and direct mailing or for broadcast media like radio and television. The advertising department decides on a name and a trademark for the product as well as a theme or motto for the campaign.

Package designers and artists match the product to the best packaging system. A good package is easy to make, adds little to the product cost, displays the best product features, and attracts consumer attention. Package designers want to make their product stand out from the crowd. They use bright colors as well as unusual package shapes, sizes, and trademarks.

Salespeople identify ways to get the manufactured product into consumers' hands. The consumer could be either an individual or another industry. When an individual buys a manufactured product in a store, it is called a retail sale. When another industry buys a manufactured product, it is called an industrial sale. The job of salespeople, whether in retail or industrial sales, is to point out the product's best features as well as how it can be sold or used.

Getting the product from the production facility to the consumer is the job of distribution or shipping workers. Distribution or shipping workers identify the most economical and efficient way to transport the product to the retail store. Comparisons of truck, train, boat, or plane transportation are made based on the cost of delivery, speed of delivery needed, and location of the retail store.

Finance Department. Financial record-keeping is very important to manufacturing companies. The main objective of a manufacturing company is to make a profit on the product they produce and sell. The finance department keeps track of all income and expenditures. Expenditures include payroll; materials costs; utility costs for lighting, heating, air conditioning, and machines; and the costs of buying and maintaining tools, equipment, and machines. Accountants, bookkeepers, and other record-keepers work in the finance department to make sure the company develops and follows a budget.

Human Resources Department. The human resources department is responsible for the following:

1. Recruitment – Recruiters identify new workers to fill vacant positions in the manufacturing company. They must match workers to the jobs by examining job applications and by conducting interviews and performance testing.

2. Training—Trainers are teachers in industry. They prepare new workers for their jobs or teach experienced workers how to use new equipment.

3. Public Relations—Public relations people keep the general public aware of the most recent developments in the company.

It is important that human resources workers get along well with other people. Their job is working with people to make the manufacturing company run smoothly and efficiently.

Words and Expressions

skilled worker 技术工人
semiskilled worker 半技术工人,半熟练工人
unskilled worker 非技术工人,做粗活的工人
marketing department 销售部门
trademark 商标,标志,品种
motto ['mɔtəu] n. 座右铭,格言,题词,标语
package ['pækidʒ] n. ;v. 包装,捆,束

salespeople ['seilz.pi:pl] n. 售货员,销售人员
retail ['rei:teil] n. ;v. 零售
accountant [ə'kauntənt] n. 会计,出纳
bookkeeper 簿记员
recruitment [ri'kru:tmənt] n. 补充,招收,充实

36. Numerical Control Software

Today, the product design process begins with computer-generated product concepts and designs, which are subjected to detailed analysis of feasibility, manufacturability, and even disposability. Traditionally, before process planning can generate a detailed plan for manufacturing a part or assembly, design tradeoffs are made. Assemblies are broken into parts. Product specifications are produced.

Quality criteria are determined to meet safety, environmental, and conformance with company and industry standards. Engineering drawings are produced as well as the bill of material. Final design decisions are made relating to styling, functional, performance, materials, tolerances, make-versus-buy, purchased parts, supplier selection, manufacturability, quality, and reliability.

In the process planning stage, tooling decisions are made. The sequence of production steps is planned with actions taken at each step and controls specified, i. e. the actions taken at each step, controls to be followed, and the state of the workpiece at each workstation. In computer-aided process planning (CAPP), an application program stores prior plans and standard sequences of manufacturing operations for families of parts coded using the group technology concept, which classifies parts based on similarity of geometric shape, manufacturing processes, or some other part characteristics.

All tools required to produce the final part or assembly are specified or designed. This includes molds, stamping and forming dies, jigs and fixtures, cutting tools, and other tooling. The tool design group typically works closely with the tool room and with suppliers to produce the necessary tools in time to meet production schedules.

Numerical control provides the operational control of a machine or machines by a series of computer-coded instructions comprising numbers, letters, and other symbols, which are translated into pulses of electrical current or other output signals that activate motors or other devices to run the machine.

With NC, machines run consistently, accurately, predictably, and essentially automatically. Quality and productivity are increased, and rework and leadtimes are reduced compared with manual operation. APT (Automatic Programming of Tools) was the first NC language.

APT was designed to function as an off-line, batch program using a mainframe computer system. Because of the computer resources needed and the expense, time-sharing was employed, a new simpler language to use developed. Eventually, interactive graphics-based NC programs using terminals and workstations were introduced to improve visualization and provide the opportunity for immediate feedback to the user. These developments remain the standard for NC programming today and are available on all computing platforms from PCs to mainframes.

NC software, accordingly, may be obtained as independent software packages, as dedicated NC turnkey systems that include a printer, plotter, and tape punch, or as components of CAD/CAM software packages that also provide design and drafting.

Interactive graphics-based NC part-programming technology increases product quality and simplifies the process by reducing setup time for lathes, mills, EDMs, and other machine tools. Graphics software lets users easily define part geometry, obtain immediate feedback, and visualize the results while changes are made quickly and efficiently.

NC packages accomplish four major functions: part description, machining strategy, post-processing, and factory communications.

To describe a part, most NC programming systems provide their own geometric modeling capability as an integral component of the system. This CAD-like frontend permits users to create parts by drawing lines, circles, arcs, and splines. All NC packages support 2D geometry creation and many have optional 3D modules. The 3D module permits the creation of complex surfaces or direct machining from the solid model will become accepted in NC.

A close relationship between the model creation and machining of that model is important. Major CAD/CAM vendors provide NC software that is integrated with the design and drafting function for just this reason. Operating from the same or a co-existent database, the NC software can directly access a model that has been created within the design module of the CAD/CAM system. This eliminates the requirement for translation of data from one format to another.

Some NC software vendors provide stand-alone systems. In this case, a file containing the definition of the model is imported from a CAD system. An IGES or DXF translation is usually used to convert the geometry from the originating system to the NC system. In other cases, the NC software can directly access the CAD database and avoid translation. In dedicated NC systems, complex NC calculations may be performed faster than for other systems and more NC-specific utilities are often available.

Currently, NC programming is usually done by individuals who have actual machining experience on the shop floor. This experience and the associated knowledge gained are critical in developing the machining strategy required to cut a part. It is particularly useful in handling unusual circumstances that often arise in machining complex parts.

Knowledge-based software systems can capture the NC programmer's knowledge and establish a set of rules to create a framework that can be used to lead the NC programmer through the development of the machining strategy. Default methodologies may be established based on materials, tolerances, surface roughness, machine tool availability, and part shape.

Knowledge-based systems may also be used to fully automate some NC programming tasks. This would provide for greater consistency among machining strategies and improve programming productivity.

Words and Expressions

subject to 在…条件下,根据,受到,经受
feasibility [fiːzə'biliti] n. 可行性,可能性,现实性
manufacturability 可制造性,工艺性
process planning 工艺设计
bill of material 材料表,材料清单,物料清单
computer-aided process planning (CAPP) 计算机辅助工艺设计
family of part 零件族
group technology 成组技术
mold [məuld] n. 铸模,模型
stamping die 冲模
forming die 成形钢模,锻模
jigs and fixtures 夹具(用以装夹工件和引导刀具的装置)
pulse [pʌls] n. 脉冲

batch program 批处理程序
mainframe 主计算机,大型计算机
time-sharing 分时(当多用户通过终端设备同时使用一台计算机时,系统把时间分成许多极短的时间片,分配给每个联机作业,由时钟控制中断,使各作业交错使用计算机)
interactive graphics-based 基于交互式图形的
software package 软件包,程序包
turnkey system 整套系统,交钥匙系统
plotter ['plɔtə] n. 绘图仪
tape punch 纸带穿孔
EDM 电加工
NC package 数控程序包
strategy ['strætidʒi] n. 策略,对策,计谋
post processing 后台处理,后加工
frontend 前端
spline [splain] n. 仿样,样条,(pl.) 仿样函数,样条函数
co-existent 同时共存的
specialty ['speʃəlti] n. 专长,专业(化),特制品,特殊产品
IGES (initial graphic exchange specification) 初始图形交换规范
DXF (data exchange format) 数据交换格式
knowledge-based 基于知识的
capture ['kæptʃə] v. 收集,吸收,吸取,记录
a set of rules 一组规则
default [di'fɔ:t] n. 系统设定(预置)值,默认(值),缺省(值);v. 默认

37. Computers in Manufacturing

  The computer is bringing manufacturing into the Information Age. This new tool, long a familiar one in business and management operations, is moving into the factory, and its advent is changing manufacturing as certainly as the steam engine changed it more than 200 years ago.

The basic metalworking processes are not likely to change fundamentally, but their organization and control definitely will.

In one respect, manufacturing could be said to be coming full circle. The first manufacturing was a cottage industry: the designer was also the manufacturer, conceiving and fabricating products one at a time. Eventually, the concept of the interchangeability of parts was developed, production was separated into specialized functions, and identical parts were produced thousands at a time.

Today, although the designer and manufacturer may not become one again, the functions are being drawn close in the movement toward an integrated manufacturing system.

It is perhaps ironic that, at a time when the market demands a high degree of product diversification, the manufacturing enterprises have to increase productivity and reduce cost. Customers are demanding high quality and diversified products for less money.

The computer is the key to each of these requirements. It is the only tool that can provide the quick reflexes, the flexibility and speed, to meet a diversified market. And it is the only tool that enables the detailed analysis and the accessibilty of accurate data necessary for the integration of the manufacturing system.

It may well be that, in the future, the computer may be essential to a company's survival. Many of today's businesses will fade away to be replaced by more productive combinations. Such more-productive combinations are superquality, superproductivity plants. The goal is to design and operate a plant that would produce 100% satisfactory parts with good productivity.

A sophisticated, competitive world is requiring that manufacturing begin to settle for more, to become itself sophisticated. To meet competition, for example, a company will have to meet the somewhat conflicting demands for greater product diversification, higher quality, improved productivity, and low prices.

The company that seeks to meet these demands will need a sophisticated tool, one that will allow it to respond quickly to customer needs while getting the most out of its manufacturing resources.

The computer is that tool.

Becoming a "superquality, superproductivity" plant requires the integration of an extremely complex system. This can be accomplished only when all elements of manufacturing—design, fabrication and assembly, quality assurance, management, materials handling—are computer integrated.

In product design, for example, interactive computer-aided-design (CAD) systems allow the drawing and analysis tasks to be performed in a fraction of the time previously required and with greater accuracy. And programs for prototype testing and evaluation further speed the design process.

In manufacturing planning, computer-aided process planning permits the selection, from thousands of possible sequences and schedules, of the optimum process.

On the shop floor, distributed intelligence in the form of microprocessors controlled machines, runs automated loading and unloading equipment, and collects data on current shop conditions.

But such isolated revolutions are not enough. What is needed is a totally automated system, linked by common software from front door to back.

The benefits range throughout the system. Essentially, computer integration provides widely and instantaneously available, accurate information, improving communication between departments, permitting tighter control, and generally enhancing the overall quality and efficiency of the entire system.

Improved communication can mean, for example, designs that are more producible. The NC programmer and the tool designer have a chance to influence the product designer, and vice versa.

Engineering changes, thus, can be reduced, and those that are required can be handled more efficiently. Not only does the computer permit them to be specified more quickly, but it also alerts subsequent users of the data to the fact that a change has been made.

The instantaneous updating of production-control data permits better planning and more effective scheduling. Expensive equipment, therefore, is used more productively, and parts move more efficiently through production, reducing work-in-process costs.

Product quality, too, can be improved. Not only are more-accurate designs produced, for example, but the use of design data by the quality-assurance department helps eliminate errors due to misunderstandings.

People are enabled to do their jobs better. By eliminating tedious calculations and paperwork—not to mention time wasted searching for information—the computer not only allows workers to be more productive but also frees them to do what only human beings can do: think creatively.

Computer integration may also lure new people into manufacturing. People are attracted because they want to work in a modern, technologically sophisticated environment.

In manufacturing engineering, CAD/CAM decreases tool design, NC-programming, and planning times while speeding the response rate, which will eventually permit in-house staff to perform work that is currently being contracted out.

Words and Expressions

cottage industry 家庭手工业

conceive [kən'si:v] v. 设想，想象，表现

interchangeability ['intə,tʃeindʒə'biliti] a. 互换性，可替换性

diversification [dai,və:sifi'keiʃən] n. 多样化，变化

coherent [kəu'hiərənt] a. 紧凑的，连贯的

coherent system 单调关联系统（指系统具有单调可靠性结构函数，且系统中任一部分的状态都与系统状态有关），相干系统

reflex ['ri:fleks] n. 反射，反应能力

accessibility [æk'sesibiliti] n. 可接近性，易维护性，检查，操作

interactive [intər'æktiv] a. 交互式的，人机对话的

distributed intelligence 分布式智能

schedule ['skedju:l] n. 时间表，进度表，预订计划

work-in-process 在制品（即在一个企业的生产过程中，正在进行加工、装配或待进一步加工、装配或待检查验收的制品）

in-house [in'haus] a. 国内的，机构内部的，公司内部的

contract out 订合同把工作包出去

38. Automated Assembly Equipment Selection

Manufacturers are facing ever-increasing customer demands and constantly rising competitive pressures. More than ever before, new products require the shortest possible time to market.

Acquiring the necessary assembly automation equipment requires a high level of risk and complexity. A combination of factors, which are difficult to package into a one-size-fits-all solution, make this process very complex. For instance, every assembly automation project is unique. Each solution or system is also unique. However, there are often many solutions that are as equally good as they are different.

Most automated assembly systems purchased today are used for assembling products that are not currently being produced or marketed. These new products will not be assembled in any significant volume prior to the installation of the system and any peripheral equipment.

Automated assembly projects typically begin by identifying a product that requires automation rather that manual assembly.

Knowledge and definition are the most critical tools manufacturng engineers need to navigate the complex maze of the equipment acquisition process. Key components include:

1. Complete working knowledge of the capabilities of those vendors from whom you plan to ask for proposals. This should include information on the types of equipment produced, industries served, company history, a top-10 customer list, details on the company's most recent installation and financial information.

2. Development in writing of as much project information as possible. This should include a project definition, project specifications and key project personnel from product design, manufacturing and purchasing.

3. Earliest possible identification of selected builders of all custom automated systems required.

4. Total commitment of the project manager and his or her team to the project. This should include allocated time for a visit to the vendor.

Before the automation procurement process begins, manufacturing engineers must have in-depth knowledge of the types of equipment produced and the capabilities of potential suppiers. There are many ways to acquire this knowledge, but a personal visit is the best.

In the early stages of projects, upper management is often reluctant to allow engineers the time and funding required to make in-person visits, believing that this can wait until after a vendor has been selected. This is flawed thinking. Travelling to prospective vendors prior to requesting proposals can avoid the enormous cost of poor vendor selection later on.

The knowledge gained from such visits greatly helps manufacturing engineers effectively define their project, compare proposals and eventually purchase a system best suited to their specific project's requirements.

Once a list of potential vendors has been identified, a certain level of project definition is required before accurate, meaningful project equipment proposals can be obtained. Specific areas that should be carefully considered include:

1. Product definition, including size and weight parameters.
2. Product drawings, including assembly and individual components.
3. System delivery requirements. This should include both system acceptance by the vendor and system installation and acceptance by the buyer.
4. Specific process or testing requirements for both components and finished, assembled products.
5. Applicable financial information, such as project budgets and return on investment (ROI) requirements.

It is much easier to define project requirements when you have first identified a well-researched list of potential equipment suppliers. The quality of the information and feedback you receive will directly affect the quality of the proposals you receive. The more complete the information, the more accurate the proposal. It's extremely helpful if some equipment description or specification is part of the request for proposal.

With a thorough knowledge of the potential builder's capabilities and areas of expertise, it usually becomes clear, as proposals are analyzed, where the best technical solution can be realized.

Obviously, system pricing must also be considered. While price is not irrelevant, it should not be among the top three considerations. Areas far more critical in the vendor selection process are:

1. The builder's ability to adequately address product process and production requirements.

2. The builder's experience and financial condition.

3. The cost of system ownership.

Overall system price is often a moving target until the project is nearly complete. But, the principles of concurrent engineering dictate that vendors of special systems be selected before final product design and before final, firm price proposals. The normal process associated with more standard capital purchases—creating a project budget and having fixed funds allocated—does not work well.

Today's globally competitive manufacturing environment dictates that real life partnerships between product manufacturers and equipment builders must be established at the earliest possible point in a project. There must be mutual respect and active, two-way dialogue from project inception to completion. Project managers must be joined at the hip to both their own product design group and the equipment builder's engineering staff.

Traditional ROI formulas rarely work with custom assembly automation. Attempts to cram the cost of special manufacturing systems into these formulas is usually unsuccessful.

Corporate purchasing poilcies should consider product life when determining ROI on custom-engineered systems. Most products that are assembled with automated equipment have life cycles of 5 years or more. Purchasing equipment because of corporate insistence that overall system costs fit 1- or 2-year ROI formulas is a big mistake.

The price of an assembly system and the cost of owning and operating it tend to be mutually exclusive. In the real world, some types of equipment outperform others. Some builders are better than others.

The goal should always be to own a system that, day in and day out, year in and year out, has the lowest amount of scheduled or unscheduled downtime and yields the maximum in net production.

When an assembly system performs at or above expectation levels, no one ever asks how much it cost. When a system performs poorly, however, people always ask how much it cost. Manufactruing engineers who are responsible for selecting a system builder should do as much homework as possible on the cost of ownership before making a final selection.

World and Expressions

acquiring *n.* 瞄准,发现
one-size-fits-all 全能的,可以符合或适用各种要求的
peripheral [pəˈrifərəl] *a.* 外围的;*n.* 外围设备,辅助设备
navigate [ˈnævigeit] *v.* 使通过,沿…航行
maze [meiz] *n.* 曲径,迷宫,迷惑
proposal 建议,方案,标书,(科研)开题报告
supplier [səˈplaiə] *n.* 供应商,厂商
in-person 亲身的,亲自的,在现场的
return on investment 投资回报率
real life 真实的,实际生活中的
mutual respect 相互尊重
inception [inˈsepʃən] *n.* 起初,开端,开始

hip 熟悉内情的,内行的,机灵的,敏感的
cram [kræm] v. 填满,勉强塞入
engineered a. 设计的
insistence [in'sistəns] n. 坚持,坚决主张
mutually exclusive 互斥的,不相容的
outperform [autpə'fɔːm] v. 做得比…好,胜过
day in and day out 天天,一天接一天
year in and year out 年年不断,始终不断地
unscheduled [ʌn'ʃedjuːld] a. 事先未安排的,计划外的
net production 纯商品生产,净产量
homework 预先的准备工作

39. Motion Control Advances Assembly

Advances in motion control and precision positioning technology have reduced the cost of many assembly automation tasks. While the most significant advances have been in electronic motion controls, the availability and lower cost of precision mechanical linear positioning and drive products have made economical automation accessible to medium and small manufacturers.

For more than a decade, microprocessor-based controllers have driven the expansion of CNC machine tools and robots. The controllers for these dedicated tools of automation are quite sophisticated. They are required to provide a broad range of motion control—from simple point-to-point to complex continuous paths for machining profiles or tracking seams for welding. Typically, 3-4 axes of motion cost $15,000-$25,000. And more importantly, the devices include many feature that are not required for automation tasks.

Initially, motion controllers for automation were marginal

substitutes for CNC type controllers. Their inherently simpler programming and lower cost made them attractive alternatives for a wide range of simple motions. The major technical advances that have dramatically expanded the availability of simpler, lower-cost, and higher-performance motion controllers came from electronic product developments.

Microprocessors have become more powerful with second- and third-generation products. These have moved from 8- bit to 16- and 16/32-bit processors that work at 12-16Mhz clock rates. Today's processors, including high-speed digital signal processors for quick arithmetic operations, have enhanced the capabilities of motion controllers significantly.

A second development paralleling microprocessor advances was power metal-oxide semiconductor, field-effect transistor (MOSFET) technology. These high-speed switching devices replaced slower, less-efficient, bipolar transistors and improved the performance of both stepping-motor drives and servo amplifiers.

Compact, and with improved power dissipation, MOSFET drives substantially improve motor performance and system response time. Stepping motors using MOSFET drives and high-performance microcontrollers operate with more torque at speeds double the 30 rev/sec once considered the maximum for this motor.

Improvements in sensor technology—especially vision—as well as new brushless and enhanced stepping motors, and intelligent peripherals, have advanced motion and positioning technology. These improvements in electronic motion control have resulted in motion controllers with features and performance historically limited to only the most expensive CNC or robot controllers.

New, low-friction anti-backlash nuts, and simple pre-loaded ball nuts, offer excellent repeatability features in both English and metric dimensions.

Advances in flexible belt technology offer high strength, reduced stretch, and improved repeatability. Widely used in Europe, belts are becoming more popular for applications requiring high speed and long travel.

Direct drives, both DC linear and high-torque motors, offer unique advantages for high-precision, high-speed, and specialty applications. Direct-drive motors have been instrumental in increasing SCARA (see Fig. 39.1) robot performance. High costs presently limit their application but are expected to drop as usage increases.

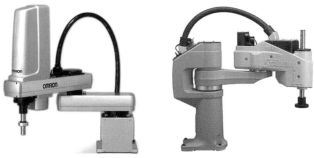

Figure 39.1   SCARA robots

Positioning Systems

Precision linear bearing positioning tables and electronic motion control products are easily integrated as a system or sub-system solution to an automation application. To achieve an economical/functional solution, mix and match the individual axes, type of drive, and motion controller to the application.

Initial integration of mechanical and electronic motion for assembly automation produced the robot—a dedicated, but universal automation tool. Robots have become more accurate, versatile, and faster as a result of improved motion controllers and machine vision technology.

Formerly, selecting a controller and drive system meant choosing either stepper equipment, which was typically the low performance and low cost option, or servo equipment, which was typically the higher cost and higher complexity option.

Today's motion controllers mix servos and steppers in a multi-axis configuration to obtain the best features of both. For a vertical axis, the high-holding torque favors the stepper. High speed and high torque requirements dictate a servo. The load-bearing capability, accuracy, and cost of the linear bearing and drive mechanism can be selected to match the mechanical features to the application requirements.

The flexibility and versatility of modular robotic or positioning systems is greatly enhanced because overall performance need not be compromised. Much of the advantage of dedicated systems, achieved through focused application engineering integration, is available to assemblers through today's motion control and precision linear products.

Cost-Effective Automation

Simple but powerful motion control and positioning products provide automation solutions at low cost and risk. Many manual assembly processes, such as fastening, soldering, testing/probing, and parts transfer involve basic requirements for precision and commercial-accuracy positioning.

Machine interface and sequencing requirements are readily handled through programming of I/O in the motion controller.

For example, a machinery supplier to domestic auto manufacturers received an order to design and manufacture an assembly line for building starter motors. Four different but similar types of motors were to be assembled with minimum machine changeover time.

Formerly, the supplier had used SCARA robots for similar applications—but cost considerations forced the company to seek other alternatives. Facility managers re-evaluated the requirements and concluded that at some stations the robot offered more axes of motion than were needed.

Ultimately, the company replaced the robot with an intelligent linear motion system that includes programmable axes and I/O. Software in the motion logic controller allows workers to calculate velocity, acceleration, and move-time required for the loads. Engineering time was substantially reduced by eliminating lengthy calculations.

Advanced motion controls and low-cost linear bearing actuators offer numerous options for both simple and complex automation. Positioning, or modular robotic systems, are simply integrated to handle a wide range of assembly applications. For the smaller user or integrator, the choices can reduce engineering time and overhead and increase productivity without sacrificing performance.

Words and Expressions

seam [si:m] n. 接缝,焊缝,接口
marginal ['mɑ:dʒinl] a. 临界的,勉强合格的
substitute ['sʌbstitju:t] v. 代替;n. 代用品,代替者
inherently [in'hiərəntli] ad. 固有地,内在地,特有地
MOSFET 金属氧化物半导体场效应晶体管
switching device 开关装置
bipolar transistor 双极性晶体管,场效应晶体管
dissipation [disi'peiʃən] n. 消耗
power dissipation 功率耗散,功率消耗
enhanced 增强的,增强型的

anti-backlash 消隙,消除间隙
ball nut 球螺母,滚珠螺母
SCARA(selectively compliant assembly robot arm) 选择性顺应组装机器人臂
linear bearing 直线轴承
load-bearing capability 承载能力
drive mechanism 驱动机构,驱动装置
starter ['stɑ:tə] n. 起动器,起动装置
solder ['sɔ:ldə] n. 低温焊料,结合物;v. 低温焊接,软(锡,银,钎)焊
engineering time 维修时间
I/O 输入/输出
overhead ['əuvəhed] n. 经常费,管理费,杂费

40. Versatility of Modular Fixturing

Today's manufacturing objectives, such as those made necessary by the just-in-time philosophy, require lead times that are shorter than ever before. Just as critical is the fact that approximately 80 percent of machining done today is in the short-run category—usually 50 to 200 pieces. This, coupled with relatively short product life and frequent design changes, can make it difficult to justify hard, or dedicated, fixturing.

On the other hand, modular fixtures can address these situations because they can be made ready at any time, within a matter of a few hours. The modular tool allows production startup of machining operations almost immediately. Also, a modular fixture can serve as a proven model to be copied if a situation warrants the need for dedicated tooling. This is not to say, however, that modular tooling will not work in a production situation; it will.

If there is a sudden increase in production demand for a product, modular tooling can serve in a production environment. It also makes a lot of sense in new product development, where there can be many component design changes, which in turn, necessitate changes in machining operations. In these cases, it is much simpler and cheaper to modify a nondedicated fixture than to rework a hard fixture.

Nondedicated fixtures can also substitute for fixtures that have either been lost or scrapped. The use of these fixtures can also relieve the burden on a toolroom and help fill the void created by the vanishing experienced toolmaker. Often, fixture-building time is critical. Many large manufacturers wait months for dedicated fixtures to be built for use in production situations. In the meantime, with revisions being made in part designs, the part may not fit the fixture when it is finally released for production. Typically, nondedicated fixturing can be built to handle almost any workpiece configuration in just a matter of hours.

Fixture components

Modular fixturing systems, such as the Halder System, available from Flexible Fixturing Systems Inc, E Granby, CT, are comprised of building-block components. The unit on which the blocks are built is the base plate.

Base plates with T-slot configuration in the X and Y axes offer positive zero-point positioning at every T-slot intersection. In the case of Halder, they are made from 5115 Series steel, case hardened to 60 RC, and ground to within 0.0004 " over 40". Four keyways on the bottom insure positive location on the machine table. Rectangular base plates are machined on two sides and have T-slots. This allows base plates to be connected to one another for additional surface when needed. T-slots on the sides also permit fixture components to be mounted to this vertical surface when needed.

The T-clamping block is used to clamp mounting blocks to the base plate. It is made of the same steel as the base plate, hardened, finish ground on two sides to ensure a precise fit, and is held together by two short screws.

In the case of this particular system, multiples of 40 basic components are supplied for a total of 247 fixture components per set.

Building a fixture

In building a modular fixture, some advice is offered by Donald H Koch, Flexible Fixturing Systems Inc. He says, "To obtain maximum benefits from nondedicated fixturing, it is recommended that a separate area be designated as a fixture assembly area. Typically, this should be a caged area free of other types of fixturing devices."

"Because nothing is perfect, there are instances when special fixture components may be required. Therefore, a toolmaker should be assigned to the task of building nondedicated fixtures. This is because training time will be nominal—probably a day or two—and, a toolmaker has skills to build any special components that may be needed, while not delaying the fixture-build sequence."

"The fixture crib should be equipped with a workbench, appropriate gages and measuring devices, and storage facilities. Some companies provide a means to hang all fixture components on the wall in full view of the toolmaker. Keep in mind the premise that modular fixtures are made up of premade, off-the-shelf components, ready to meet machining objectives."

Storage and reuse

The premise of nondedicated fixturing is that it can be used over and over again. Construction details of a fixture must be detailed and recorded, however, for reuse at a later date. After the fixture has been built, a photograph is taken. All components used in construction are listed on a preprinted assembly card. And, a drawing is made showing the exact location of each element used.

When the production run is completed, the fixture is disassembled and the assembly data card is filed for future reference. If and when the fixture is needed again, the coordinate grid method of recording the fixture position on the base plate assures that it can be reassembled with assurance of a high degree of accuracy. Another form of documentation—CAD 3-D software—is available in IGES and CATIA formats. Once stored in a computer's memory, the fixture may be reproduced at any time.

According to Don Koch, "There is a valid statistic that says you can expect to spend five times the purchase price of a machine tool on dedicated fixturing over the useful life of the machine. The same machine tool can be tooled with modular fixturing at a lower single investment, and can then be reused over the life of the machine tool."

Words and Expressions

versatility [və:sə′tiliti] n. 通用性,多功能性,多方面适用性
modular fixturing 组合夹具
coupled with 加上,外加
address v. 专注于,致力于,从事于
proven a. 被证实的,可靠的
scrapped 废弃的
vanishing [′væniʃiŋ] n. 消失
toolmaker 工具制造工
base plate 基础板,底板
intersection [intə′sekʃən] n. 相交,十字路口,交叉点

case harden 表面硬化,表面淬火
rectangular [rek'tæŋgjulə] a. 矩形的,成直角的
fit 配合
multiple ['mʌltipl] a. 多样的;n. 倍数,若干 v. 成倍增加
nominal ['nɔminl] a. 极小的,按计划进行的
fixture crib 夹具室
be equipped with M 装备有 M,安装有 M
workbench 工作台
premise ['premis] n. 前提,前言,根据
premade 预先做的
off-the-shelf 成品的,畅销的,现成的,流行的
disassemble [disə'sembl] v. 拆卸,拆除,分解
coordinate grid 坐标网
CATIA (computer aided three dimensional interactive application) 计算机辅助三维交互式应用
tool [tu:l] n. 工具;v. 给…装备上工具,提供加工机械

41. Product Design for Manufacture and Assembly

Design for manufacture (DFM) means different things to different people. For the individual whose task is to consider the design of a single component, DFM means the avoidance of component features that are unnecessarily expensive to produce. Examples include the following:

• Specification of surfaces that are smoother than necessary on a machined component, necessitating additional finishing operations

• Specification of wide variations in the wall thickness of an injection-molded component

· Specification of too-small fillet radii in a forged component

· Specification of internal apertures too close to the bend line of a sheet metal component.

Alternatively, the DFM of a single component might involve minimizing material costs or making the optimum choice of materials and processes to achieve a particular result. For example, can the component be cold-headed and finish-machined rather than machined from bar stock? All of these considerations are important and can affect the cost of manufacture. They represent only the fine-tuning of costs, however, and by the time such considerations are made, the opportunities for significant savings may have been lost.

It is important to differentiate between component or part DFM and product DFM. The former represents only the fine-tuning process undertaken once the product form has been decided upon; the latter attacks the fundamental problem of the effect of product structure on total manufacturing costs.

The key to successful product DFM is product simplification through design for assembly (DFA). DFA techniques primarily aim to simplify the product structure so that assembly costs are reduced. Experience is showing, however, that the consequent reductions in part costs often far outweigh the assembly cost reductions. Even more important, the elimination of parts as a result of DFA has several secondary benefits more difficult to quantify, such as improved reliability and reduction in inventory and production-control costs. DFA, therefore, means much more than design to reduce assembly costs and, in fact, is central to the issue of product DFM. In other words, part DFM is the icing on the cake; product DFM through DFA is the cake.

DFA derives its name from a recognition of the need to consider assembly problems at the early stages of design; it therefore entails the analysis of both product and part design. For some years now, an assembly evaluation method (AEM) has been in use at Hitachi. In this proprietary method, commonly referred to as the Hitachi method, assembly element symbols are selected from a small array of possible choices. Combinations of the symbols then represent the complete assembly operation for a particular part. Penalty points associated with each symbol are substituted into an equation, resulting in a numerical rating for the design. The higher the rating, the better the design.

Another quantitative method, known as the Boothroyd and Dewhurst method, involves two principal steps:

1. The application of criteria to each part to determine whether, theoretically, it should be separate from all the other parts in the assembly.

2. An estimate of the handling and assembly costs for each part using the appropriate assembly process—manual, robotic, or high-speed automatic.

The first step, which involves minimizing the part count, is the most important. It guides the designer toward the kind of product simplification that can result in substantial savings in product costs. It also provides a basis for measuring the quality of a design from an assembly standpoint. During the second step, cost figures are generated that allow the designer to judge whether suggested design changes will result in meaningful savings in assembly cost.

The third quantitative method used in industry is the GE/Hitachi method, which is basically the Hitachi method with the Boothroyd and Dewhurst criteria for part-count reduction added.

For business reasons, companies are seldom prepared to release their manufacturing cost information. One reason is that many companies are not sufficiently confident about their costing procedures to want manufacturing costs made public for general discussion. In such an environment, designers will often not be informed of the cost of manufacturing the product they have been designing. Moreover, designers do not usually have the tools necessary to obtain immediate cost estimates relating to alternative product design schemes. Typically, a product will have been designed and detailed and a prototype manufactured before a manufacturing cost estimate is attempted. Unfortunately, by then it is too late. The opportunity to consider radically different product structures has been lost, and among those design alternatives might have been a version that is substantially less expensive to produce.

Currently, there is much interest in having product DFM and DFA techniques available on CAD/CAM systems. By the time a proposed product design has been sufficiently detailed to enter it into the CAD/CAM system, however, it is already too late to make radical changes. A CAD representation of a new product is an excellent vehicle for making effortless detail changes, such as moving holes and changing draft angles. But for considering product structure alternatives, such as the choice of several machined parts versus one die casting, a CAD system is not nearly as useful. These basic, fundamentally important decisions must be made at the early sketch stage in product design.

A conflict thus exists. On the one hand, the designer needs cost estimates as a basis for making sound decisions; on the other hand, the product design is not sufficiently firm to allow estimates to be made using currently available techniques. The means of overcoming this dilemma is another key to successful product DFM—namely, early cost estimating.

A DFM system must therefore be able to predict both assembly costs and component manufacturing costs at the earliest stages of product design. Only in this way will it be possible to design a product that takes maximum advantage of the capabilities of chosen manufacturing processes within the constraints imposed by anticipated production volumes. In many situations, this will simply mean providing designers or design teams with the software tools that will enable them to make sound judgments from a range of choices. These choices may involve designs necessitating increased tooling costs but fewer different parts and reduced assembly costs.

It is anticipated that the product DFM considerations will always start with DFA. To aid designers in implementing these techniques, the authors have developed DFA software that allows a designer to establish an efficient assembly sequence for a proposed new product concept. The software then questions the relationship between the parts and gives an assembly efficiency rating, together with estimated assembly costs.

The DFA process uses the assembly sequence as a vehicle for analyzing the product structure in order to force the design toward more integrated solutions with a reduced part count. This result of DFA is often the most important one in achieving total product cost reductions. Thus, DFA analyses must be supported by techniques that will allow the design team to make early estimates of material, processing, and tooling costs. Only in this way can different designs, with different numbers of parts and perhaps using different materials and processes, be compared before a detailed design commitment is made.

The techniques of DFA and DFM can play a major role in reducing costs and increasing productivity. Recognition of this fact is also increasing the demand for cost-estimating tools that allow design teams to make the necessary tradeoffs at the early concept stages of design.

Words and Expressions

design for manufacture(DFM) 面向制造的设计
injection-molded 注射成型的,注塑成型的
fillet ['filit] n. 圆角,倒角
bend [bend] n.;v. 弯曲,弯曲部
cold-headed 冷镦的
fine-tuning 微调,细调
secondary ['sekəndəri] a. 次要的,从属的,第二位的,间接的
secondary benefit 间接利益,次级收益
quantify ['kwɔntifai] v. 确定数量,量化
central ['sentrəl] a. 中央的,中心的,重要的,主要的
icing ['aisiŋ] n. (糕饼表层的)糖霜,酥皮
derive [di'raiv] v. 起源于,来自
entail [in'teil] v. 需要,要求,使…发生,带来
Hitachi [hi'tɑ:tʃi] n. 日立
proprietary [prə'praiətəri] a. 专利的,专有的,独占的
an array of 一排,一系列
automatic [ɔ:tə'mætik] n. 自动机械,自动装置
detail v. 清晰地说明;详细叙述
vehicle n. 车辆,媒介物,运载工具,载体
radically ['rædikəli] ad. 根本地,安全地,主要地
effortless ['efətlis] a. 容易的,不费力气的
draft angle 拔模斜度
dilemma [di'lemə] n. 困境,进退两难
cost estimating 费用概算,成本估计
depression [di'preʃən] n. 沉降,凹陷
molded part 模制零件
metallurgy [me'tælədʒi] n. 冶金学,冶金

powder metallurgy 粉末冶金

42. Making a Cost Estimate

In many companies cost estimating is accomplished by a professional who specializes in determining the cost of a component whether it's made in-house or purchased from an external source. This person must be as accurate as possible in his or her estimates, as major decisions about the product are based on these costs. Unfortunately, cost estimators need fairly detailed information to perform their job. It is unrealistic for the designer to give the cost estimator 20 conceptual designs in the form of rough sketches and expect any cooperation in return. Thus, it is essential that the designer be able to make at least rough cost estimations until the design is refined enough to seek a cost estimator's aid. (In many small companies, all cost estimations are done by the designer.)

The first estimations need to be made early in the product design phase. These must be precise enough to be of use in making decisions about which designs to eliminate from consideration and which designs to continue refining. At this stage of the process, cost estimates within 30 percent of the final direct cost are possible. The goal is to have the accuracy of this estimate improve as the design is refined toward the final product. The more experience in estimating similar products, the more accurate the early estimates will be.

The cost estimating procedure is dependent on the source of the components in a product. There are three possible options for obtaining the components: (1) purchase finished components from a vendor; (2) have a vendor produce components designed in-house; or (3) manufacture components in-house.

There are strong incentives to buy existing components from vendors. If the quantity to be purchased is large enough, most vendors will work with the product designer and modify existing components to meet the needs of the new product.

If existing components or modified components are not available off the shelf, then they must be produced, in which case the decision must be made as to whether they should be produced by a vendor or made in-house. This is the classic "make or buy" decision, a complex decision that must be based not only on the cost of the component involved, but on the capitalization of equipment, the investment in manufacturing personnel, and plans by the company to use similar manufacturing equipment in the future.

Regardless of whether the component is to be made or bought, there is a need to develop a cost estimate. We look now at cost estimate for machined components.

Machined components are manufactured by removing portions of the material not wanted. Thus the costs for machining are primarily dependent on the cost and shape of the stock material, the amount and shape of material that needs to be removed, and how accurately it must be removed. These three areas can be further decomposed into six significant factors that determine the cost of a machined component:

1. From what material is the component to be machined? The material affects the cost in three ways: the cost of the raw material, the value of the scrap produced, and the ease with which the material can be machined. The first two are direct material costs and the last affects the amount of labor needed, the time and machines that will be tied up manufacturing the component.

2. What type of machine will be used to manufacture the component? The type of machine—lathe, horizontal mill, vertical mill, etc. to be used in manufacture affects the cost of the component. For

each type, there is not only the cost of the machine time itself, there is also the cost of the tools and fixtures needed.

3. What are the major dimensions of the component? This factor helps determine what size machines of each type will be required to manufacture the component. Each machine in a manufacturing facility has a different cost for use, depending on the initial cost of the machine and its age.

4. How many machined surfaces are there and how much material is to be removed? Just knowing the number of surfaces and material removal ratio (the ratio of the final component volume to the initial volume) can aid in giving a good estimate for the amount of time required to machine the part. More accurate estimates require knowing exactly what machining operations will be used to make each cut.

5. How many components will be made? The number of components to be made has a great effect on the cost. For one piece, fixturing will be minimal, though long setup and alignment times will be required. For a few pieces, some fixtures will be made. For a high volume, the manufacturing process will be automated, with extensive fixturing and numerical controlled machining.

6. What tolerance and surface roughness are required? The tighter the tolerance and surface roughness requirements, the more time and equipment needed in manufacture.

Words and Expressions

estimate ['estimeit] *v.* ; *n.* 估计, 估价, 预测
in return 作为回报, 替换
in-house 机构内部的, 自身的
sketch [sketʃ] *n.* 略图, 草图, 概略, 拟定; *v.* 绘略图
vendor ['vendɔ:] *n.* 卖主, 供货商

scrap [skræp] n. 废料,残余物,边角料
fixture ['fikstʃə] n. 夹具,夹紧装置
facility [fə'siliti] n. 设备,工具

43. Determining the Cost of a Product

The total cost of a product to the customer, the list price, is shown in Fig. 43.1 broken down into its constituent parts. All costs can be lumped into two broad categories, direct costs and indirect costs. *Direct costs* are those that can be traced directly to a specific component, assembly, or product. All other costs are called *indirect costs*. The terminology generally used to describe the differing costs that contribute to the direct and indirect costs is defined below. Each company has its own method of bookkeeping, so the definitions given here may not match every accounting scheme. However, every company needs to account for all the costs discussed, whatever the terminology used.

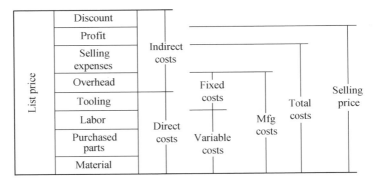

Figure 43.1   Product cost breakdown

A major part of the direct cost is the *material costs*. This includes the expenses of all the materials that are purchased for a product, including the expense of the waste caused by scrap and spoilage. Scrap is often an important consideration. For most materials the scrap can be reclaimed, and the return from the reclamation can be deducted from the material costs. Spoilage includes parts and material that may not be usable because of deterioration or damage.

Components that are not fabricated, but purchased from vendors, are also considered as direct costs. At a minimum, this *purchased-parts cost* will include the cost of fasteners and maybe the packaging to ship the product. At a maximum, all components may be made outside the company and only assembly performed in house. In this case there are no material costs.

The *labor cost* is the cost of wages and benefits to the work force needed to manufacture and/or assemble the products. This value must include not only the employees' salaries, but all fringe benefits, including medical insurance, retirement funds, and vacation times. Additionally, some companies include overhead (to be defined below) in figuring the direct labor cost. With fringe benefits and overhead included, the labor cost will be two to three times the salary of the worker.

The last element of direct costs is the *tooling cost*. This cost includes all jigs, fixtures, molds, and other parts specifically manufactured or purchased for the production of the product.

Figure 43.1 shows that the sum of the material, labor, purchased parts, and tooling used is the direct cost. The *manufacturing cost* is the direct cost plus the *overhead*, which includes all administration, engineering, secretarial, cleaning, rent or lease on building, utilities, and other costs that occur day to day, even if no product rolls out the door. Some companies subdivide the overhead into engineering overhead and administrative overhead, where engineering portion includes all the expenses associated with research, development, and design of the product.

The manufacturing cost can be broken down in another important way. The material, labor, and purchased-parts costs are *variable costs*, as they vary directly with the number of units produced. For most high-volume processes, this variation is nearly linear: It costs about twice as much to produce twice as many units. However, at lower volumes the costs may change drastically with volume. Other manufacturing costs such as tooling and overhead are *fixed costs*, as they remain the same regardless of the number of units made. Even if production fell to zero, funds spent on tooling and the expenses associated with the facilities and nonproduction labor would still remain the same.

The *total cost* of the product is the manufacturing cost plus the selling expenses. It accounts for all the expenses needed to get the product to the point of sale. The actual *selling price* is the total cost plus the *profit*. Lastly, if the product has been sold to a distributor or a retail store (anything other than direct sales), then the actual price to the consumer, the list price, is the selling price plus the *discount*. Thus the discount is the part of the list price that covers the costs and profits of retail sales. If the design effort is on a manufacturing machine to be used in-house, then costs such as discount and selling expenses do not exist, but depending on the bookkeeping practices of the particular company, there may still be profit included in the cost.

The salaries for the designers, drafter, and engineers, and the costs for their equipment and facilities are all part of the overhead. Designers have little control over these fixed expenses, beyond using their time and equipment efficiently. The designer's big impact is on the direct costs: tooling, labor, material, and purchased-parts costs. The cost of design and its influence on the manufacturing cost can be seen in Fig. 43.2, which is based on Ford Motor Company data. This data shows the manufacturing cost in the left column and the influence on manufacturing cost in the right. If it is assumed that the costs for purchased parts and tooling are subsumed in the material costs, then these account for about 50 percent of the manufacturing costs. The labor is about 15 percent, and the overhead, including design expenses, is 35 percent. As a rule of thumb, for companies whose products are manufactured mainly in-house and in high volume, the manufacturing cost is approximately three times the cost of the materials. Also, the selling price is approximately nine times the material costs, or three times the manufacturing cost. This is sometimes called the material-manufacturing-selling 1-3-9 rule. This ratio varies greatly from product to product but does give a feel for how the selling price is related to costs of materials. The Ford data in Fig. 43.2 shows a 1:2 ratio between materials cost and manufacturing cost, less than the rule would predict.

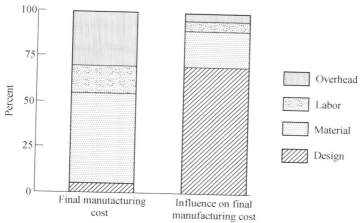

Figure 43.2   Design influence on manufacturing cost.

The right column in Fig. 43.2 shows the influence of the design effort on the manufacturing cost. As mentioned above, the designer can influence all the direct costs in a product, including the types of materials used, the purchased parts specified, the production methods, and thus the labor hours and the cost of tooling. Management, on the other hand, has much less influence on the manufacturing costs. They can negotiate for lower prices on a material specified by the designer, lower wages for the workers, or try to trim overhead. With these considerations, it is not surprising that Ford data shows 70 per cent of the influence on the manufacturing cost controlled by design.

Words and Expressions

list cost 定价,价目表所列之价格,标价
constituent [kɔn'stitjuənt] a. 组成的 n. 组成部分,构成
lump [lʌmp] v. 把…合在一起,总结,概括
direct costs and indirect costs 直接成本与间接成本
terminology [tə:mi'nɔlədʒi] n. 术语,专门名词
bookkeeping ['bukki:piŋ] n. 簿记,记账
spoilage ['spɔilidʒ] n. 损坏,变坏,废品,浪费的东西
reclaim [ri'kleim] v. 回收,再生,重复使用
reclamation [reklə'meiʃən] n. 废料回收,再生,修整
deterioration [ditiəriə'reiʃən] n. 变质,退化,恶化,变坏
packaging ['pækidʒiŋ] n. 打包,装箱,包装
fringe [frindʒ] n. 边缘;a. 边缘的,附加的,较次要的
fringe benefit 附加福利,补贴,福利金
variable cost 可变成本
fixed cost 固定成本
lease [li:s] n. 租约,租借权,租借期限;v. 租借,出租
roll out 离开
subdivide ['sʌbdi'vaid] v. 细分,再划分,重分
discount ['diskaunt] n. 折扣 v. 打折

drafter [drɑːʃtə] n. 制图者,描图者,制图机械
assume [ə'sjuːm] v. 假定,设想,采取
subsume [səb'sjuːm] v. 包含,包括,把……归入(某一类)
trim [trim] v. ;n. 使整齐,修整,调整,去毛刺

5 MODERN MANUFACTURING ENGINEERING AND DEVELOPMENT

44. Industrial Robots

There are a variety of definitions of the term industrial robot. Depending on the definition used, the number of industrial robot installations worldwide varies widely. Numerous single-purpose machines are used in manufacturing plants that might appear to be robots. These machines can only perform a single function and cannot be reprogrammed to perform a different function. Such single-purpose machines do not fit the definition for industrial robots that is becoming widely accepted.

An industrial robot is defined by the International Organization for Standardization (ISO) as an automatically controlled, reprogrammable, multipurpose manipulator, which may be either fixed in place or mobile for use in industrial automation applications.

There exist several other definitions too, given by other societies, e. g., by the Robot Institute of America (RIA), and others. The definition developed by RIA is:

A robot is a reprogrammable multifunctional manipulator designed to move material, parts, tools, or specialized devices through variable programmed motions for the performance of a variety of tasks.

All definitions have two points in common. They all contain the words reprogrammable and multifunctional. It is these two characteristics that separate the true industrial robot from the various single-purpose machines used in modern manufacturing firms.

The term "reprogrammable" implies two things: The robot operates according to a written program, and this program can be rewritten to accommodate a variety of manufacturing tasks.

The term "multifunctional" means that the robot can, through reprogramming and the use of different end-effectors, perform a number of different manufacturing tasks. Definitions written around these two critical characteristics have become the accepted definitions among manufacturing professionals.

The first articulated arm came about in 1951 and was used by the U. S. Atomic Energy Commission. In 1954, the first industrial robot was designed by George C. Devol. It was an unsophisticated programmable materials handling machine.

The first commercially produced robot was developed in 1959. In 1962, the first industrial robot to be used on a production line was installed in the General Motors Corporation. It was used to lift red-hot door handles and other such car parts from die casting machines in an automobile factory in New Jersey, USA. Its most distinctive feature was a gripper that eliminated the need for man to touch car parts just made from molten metal. It had five degrees of freedom (DOF). This robot was produced by Unimation.

A major step forward in robot control occurred in 1973 with the development of the T^3 industrial robot by Cincinnati Milacron. The T^3 robot was the first commercially produced industrial robot controlled by a minicomputer. Figure 44. 1 shows a T^3 robot with all the motions indicated, it is also called jointed-spherical robot.

Since then robotics has evolved in a multitude of directions, starting from using them in welding, painting, in assembly, machine tool loading and unloading, to inspection.

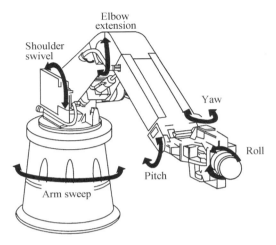

Figure 44.1 Jointed-spherical robot

Over the last three decades automobile factories have become dominated by robots. A typical factory contains hundreds of industrial robots working on fully automated production lines (see Fig. 44.2). For example, on an automated production line, a vehicle chassis on a conveyor is welded, painted and finally assembled at a sequence of robot stations.

Figure 44.2 Robots used in car factories

Mass-produced printed circuit boards (PCBs) are almost exclusively assembled by pick-and-place robots, typically with SCARA manipulators (see also Fig. 39.1), which pick tiny electronic components, and place them on to PCBs (see Fig. 44.3) with great accuracy. Such robots can place tens of thousands of components per hour, far surpassing a human in speed, accuracy, and reliability

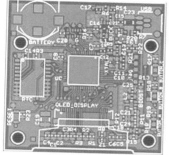 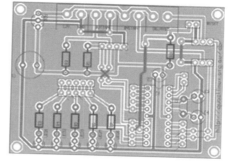

Figure 44.3 Printed circuit boards

A major reason for the growth in the use of industrial robots is their declining cost. Since 1970s, the rapid inflation of wages has tremendously increased the personnel costs of manufacturing firms. In order to survive, manufacturers were forced to consider any technological developments that could help improve productivity. It became imperative to produce better products at lower costs in order to be competitive in the global market economy. Other factors such as the need to find better ways of performing dangerous manufacturing tasks contributed to the development of industrial robots. However, the fundamental reason has always been, and is still, improved productivity.

One of the principal advantages of robots is that they can be used in settings that are dangerous to humans. Welding and parting are examples of applications where robots can be used more safely than humans. Most industrial robots of today are designed to work in environments which are not safe and very difficult for human workers. For example, a robot can be designed to handle a very hot or very cold object that the human hand cannot handle safely.

Even though robots are closely associated with safety in the workplace, they can, in themselves, be dangerous. Robots and robot cells must be carefully designed and configured so that they do not endanger human workers and other machines. Robot workspaces should be accurately calculated and a danger zone surrounding the workspace clearly marked off. Barriers can be used to keep human workers out of a robot's workspace. Even with such precautions it is still a good idea to have an automatic shutdown system in situations where robots are used. Such a system should have the capacity to sense the need for an automatic shutdown of operations.

Words and Expressions

installation [ˌɪnstəˈleiʃən] n. 整套装置, 设备, 结构, 安装
multifunctional a. 多功能的
manipulator [məˈnipjuleitə] n. 机械手, 机器人机械本体, 操作机, 操作臂
end effector 末端执行器, 末端操作器
articulated [ɑːˈtikjulitid] a. 关节式的, 铰链的
George C. Devol 乔治 C. 德沃尔(1912-2011) 美国发明家
General Motors Corporation (美国)通用汽车公司
die casting 压力铸造, 压铸
die casting machine 压铸机(在压力作用下把熔融金属液压射到模具中冷却成型, 开模后得到固体金属铸件的铸造机械)
gripper [ˈɡripə] n. 手爪, 夹爪, 夹持器
Unimation 万能自动化公司(Universal Automation)
T^3 为 The Tomorrow Tool 的缩写, 也可以写为 T3 或 T-3
jointed-spherical robot 关节式球面机器人
swivel [ˈswivl] n.; v. 旋转
yaw [jɔː] n. 侧摆, 偏摆, 偏转, 摆动

pitch [pitʃ] n. 俯仰
roll [rəul] n. 侧滚,翻转,回转
chassis ['ʃæsi] n. 底盘
mass-produced 大批量生产的
printed circuit board 印刷电路板
pick-and-place robot 抓-放型机器人,抓放机器人,取-放型机器人
SCARA 选择顺应性装配机器手臂,平面关节型装配机器人
tens of thousands 成千上万,好几万
setting ['setiŋ] n. 位置,安装,环境
parting ['pɑːtiŋ] a. 分离的,离别的;n. 分离,切断

45. Robotic Sensors

Sensors in robots are like our eyes, nose, ears, mouth, and skin. Based on the function of human organs, for example the eyes or skin etc., terms like vision, tactile etc., have been used for robot sensors. Robots like humans, must gather extensive information about their environment in order to function effectively. They must pick-up an object and know it has been picked up. As the robot arm moves through the 3-dimensional space, it must avoid obstacles and approach items to be handled at a controlled speed. Some objects are heavy, others are fragile, and others are too hot to handle. These characteristics of objects and the environment must be recognized, and fed into the computer that controls a robot's movements. For example, to move the end effector of a robot along a desired trajectory and to exert a desired force on an object, the end effector and sensors must work in coordination with the robot controller. Robot sensors can be classified into two categories: internal sensors and external sensors.

Internal sensors, as the name suggests, are used to measure the internal state of a robot, i. e. its position, velocity, acceleration, etc. at a particular instant. Depending on the various quantities it measures, a sensor is termed as the position, velocity, acceleration, or force sensor.

Position sensors measure the position of each joint of a robot. There are several types of position sensors, e. g. encoder, LVDT, etc.

Encoder is an optical-electrical device that converts motion into a sequence of digital pulses. Encoders can be either absolute or incremental type. Further, each type may be again linear or rotary.

The linear variable differential transformer (LVDT) is one of the most used displacement transducers, particularly when high accuracy is needed. LVDT are made up of three coils, as shown in Fig. 45.1. AC supplied to the input coil at a specific voltage E_i generates a total output voltage across the secondary coils. The output voltage is a linear function of the displacement of a movable ferrous core inside the coils.

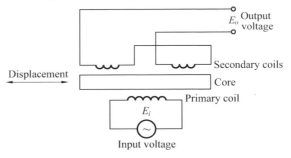

Figure 45.1　Schematic diagram of an LVDT

Electrical resistance strain gages are widely used to measure strains due to force or torque. Gages are made of electrical conductors, usually wire or foil, as shown in Fig. 45.2. They are glued on the surfaces where strains are to be measured. The strains cause changes in the resistance of the strain gages, which are measured by attaching them to the Wheatstone bridge circuit as one of the four resistances, $R_1 \cdots R_4$ of Fig. 45.3a. It is a cheap and accurate method of measuring strain. But care should be taken for the temperature changes. In order to enhance the output voltage and eliminate resistance changes due to the change in temperature, two strain gages are used, as shown in Fig. 45.3b, to measure the force at the end of the cantilever beam.

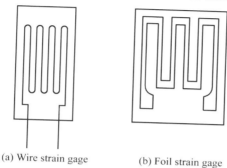

(a) Wire strain gage　　(b) Foil strain gage

Figure 45.2　Strain gages

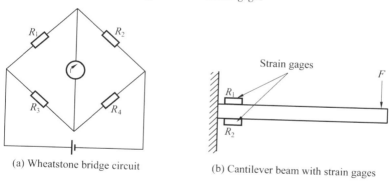

(a) Wheatstone bridge circuit　　(b) Cantilever beam with strain gages

Figure 45.3　Strain measurement

External sensors are primarily used to learn more about the robot's environment. External sensors can be divided into two categories: contact type (e.g. limit switch) and non-contact type (e.g. proximity sensor and vision sensor).

A limit switch is constructed much as the ordinary light switch used at home and office. It has the same on/off characteristics. Limit switches (see Fig. 45.4) are used in robots to detect the extreme positions of the motions, where the link reaching an extreme position switches off the corresponding actuator, thus safeguarding any possible damage to the mechanical structure of the robot arm.

Figure 45.4   Limit switches

A proximity sensor senses and indicates the presence of an object within a fixed space near the sensor. Proximity sensors are of two types, capacitive and inductive.

Vision systems are used with robots to let them look around and find the parts for picking and placing them at appropriate locations. Most current systems use cameras based on the charged coupled devices (CCDs) techniques. These cameras are smaller, last longer, and have less inherent image distortion than conventional robot cameras.

Words and Expressions

human organs 人体器官
tactile ['tæktail] *a.* 触觉的,有触觉的
fed into 被送入,被输入
trajectory [trə'dʒekətəri] *n.* 轨迹
located at 位于,设在

desired trajectory 预定轨迹,期望轨迹
in coordination with 与……配合
internal sensor 内部传感器
external sensor 外部传感器
as the name suggests 顾名思义
at a particular instant 在某个具体瞬间
force sensor 力觉传感器,力传感器
encoder [in'kəudə] n. 编码器,光电编码器
LVDT 线性可变差动变压器(Linear Variable Differential Transformer)
a sequence of 一系列的,一连串的
digital pulse 数字脉冲
absolute or incremental type 绝对式或增量式
linear or rotary 直线或旋转
most used 用得最多的
displacement transducer 位移传感器
coil [kɔil] n. 线圈
secondary coil 次级线圈
primary coil 初级线圈
ferrous core 铁芯
electrical resistance strain gage 电阻应变片
electrical conductor 导电体,电导体,电导线
attach to 连接到,加入,把……放在
Wheatstone bridge circuit 惠斯登电桥电路
care should be taken 特别要注意
wire strain gage 电阻丝式应变片(也可以写作 wire gage)
foil strain gage 箔式应变片(也可以写作 foil gage)
limit switch 限位开关,行程开关
light switch 照明开关 电灯开关
proximity [prɔk'simiti] n. 接近
proximity sensor 接近觉传感器

capacitive [kə'pæsitiv] *a.* 电容的
inductive [in'dʌktiv] *a.* 电感的，感应的
current system 现行系统
charged coupled device (CCD) 电荷耦合元件
image distortion 图像失真

46. Types of Robots

Ask a number of people to describe a robot and most of them will answer they look like a human. Interestingly a robot that looks like a human is probably the most difficult robot to make. It is usually a waste of time and not the most sensible thing to model a robot after a human being. A robot needs to be above all functional and designed with quality that suits its primary tasks. It depends on the task at hand whether the robot is big, small, able to move or nailed to the ground. Robots may be classified as follows:

Mobile Robots

Mobile robots are able to move, usually they perform task such as search areas. An example is the Mars Explorer, specifically designed to roam the mars surface.

Mobile robots are used for task where people cannot go, either because it is too dangerous or because people cannot reach the area that needs to be searched.

Mobile robots can be divided in two categories:

1. Rolling Robots: Rolling robots have wheels to move around. These are the type of robots that can quickly and easily search move around. However they are only useful in flat areas, rocky terrains give them a hard time. Flat terrains are their territory.

2. Walking Robots: Robots on legs are usually brought in when

the terrain is rocky and difficult to enter with wheels. Robots have a hard time shifting balance and keep them from tumbling. That's why most walking robots have at least 4 legs, usually they have 6 legs or more. Even when they lift one or more legs they still keep their balance. Development of legged robots is often modeled after insects or crayfish.

Stationary Robots

Robots are not only used to explore areas or imitate a human being. Most robots perform repeating tasks without ever moving a centimeter. Most robots are 'working' in industry settings. Especially dull and repeating tasks are suitable for robots. A robot never gets tired, it will perform its duty day and night without ever complaining. In case the tasks at hand are done, the robots will be reprogrammed to perform other tasks.

Autonomous Robots

Autonomous robots are self supporting or in other words self contained. In a way they rely on their own 'brains'.

Autonomous robots run a program that gives them the opportunity to decide on the action to perform depending on their surroundings. At times these robots even learn new behavior. They start out with a short routine and adapt this routine to be more successful at the task they perform. The most successful routine will be repeated as such their behavior is shaped. Autonomous robots can learn to walk or avoid obstacles they find in their way. Think about a six legged robot, at first the legs move at random, after a little while the robot adjusts its program and performs a pattern which enables it to move in a direction.

Remote-control Robots

At present, an autonomous robot is not a very clever or intelligent unit. The memory and brain capacity is usually limited, an autonomous robot can be compared to an insect in that respect. In case a robot needs to perform more complicated yet undetermined tasks an autonomous robot is not the right choice.

Complicated tasks are still best performed by human beings with real brainpower. A person can guide a robot by remote control. A person can perform difficult and usually dangerous tasks without being at the spot where the tasks are performed. To detonate a bomb it is safer to send the robot to the danger area.

Virtual Robots

Virtual robots don't exits in real life. Virtual robots are just programs, building blocks of software inside a computer. A virtual robot can simulate a real robot or just perform a repeating task. A special kind of robot is a robot that searches the World Wide Web. The internet has countless robots crawling from site to site. These Web Crawlers collect information on websites and send this information to the search engines.

Another popular virtual robot is the chatterbot. These robots simulate conversations with users of the internet. One of the first chatterbots was ELIZA. There are many varieties of chatterbots now.

Words and Expressions

above all 尤其是,特别是,最主要的是
Mars Explorer 火星探测器
roam ['rəum] v. ;n. . 漫步,漫游,徘徊
flat terrain 平地,平坦地带
tumbling ['tʌmbliŋ] n. 翻倒,滚转
crayfish ['kreifiʃ] n. 小龙虾
autonomous robot 自主机器人
self supporting 自己支持的

self contained 自持的,整装的,设备齐全的
virtual robot 虚拟机器人
building block 程序块（在软件工程中,指较高一级程序或模块使用的一个单元或模块）
crawling ['krɔːliŋ] n. 爬行
Web Crawler 网页爬行器
search engine 搜索引擎
chatterbot 聊天机器人

47. Coordinate Measuring Machines

A coordinate measuring machine (CMM) is typically used to generate 3-D points from the surface of a part. It's digitizing a part in three dimensions. However, it is often used to make 2-D measurements such as measuring the center and radius of a circle in a plane, or even one-dimensional measurements such as determining the distance between two points. Typically, CMMs are configured to measure in Cartesian coordinates. There are also CMMs that measure in cylindrical or spherical coordinates. They can measure any part surface they can reach.

CMMs typically generate points in two ways: point-to-point mode, where the CMM touches the part and generates a single point of data every time contacting with the part, or scanning, where the CMM moves over a part, generating data as it moves. Scanning generates significantly more data than contacting, but is typically not as accurate.

CMMs are manual or automatic. In manual mode, the CMM is moved by the user. An automatic CMM is typically actuated by electric drives (using ballscrews or linear motors). As shown in Figure 47.1, articulated arm CMMs look very much like six-degree-of-freedom robots, and are almost always manually driven. Hybrid CMMs are a cross between articulated arm systems and traditional CMMs. They may have servo assist for making measurements.

While the CMM hardware generates the coordinate data, the software bundled with the CMM (or in many instances sold separately) analyzes the data and presents the results to the user in a form that permits an understanding of part quality, and conformance to specified geometry.

Figure 47.1 Articulated arm CMM

New user-friendly software that allows the CMM and probe to be accurately, quickly, and easily calibrated has also made the CMM more accurate and easier to use.

The most important advancement in CMM technology over the past several years is that significant errors can be corrected mathematically via software. As a result, looser tolerances can be used on the system hardware, and the resulting errors (as long as they are highly repeatable) are eliminated in software. This results in lower manufacturing costs, while retaining or even improving the capabilities of the CMM. Other major design innovations in the past were linear air bearings and linear scales for improved repeatability and accuracy.

Examples of geometries that are difficult to measure include very deep holes, where a probe must be inserted down the length of the hole. If the hole diameter is small, such as cooling holes on turbine blades, the task becomes even more formidable.

A controlled environment is important for efficient CMM operation. CMMs can operate well on the shop floor if they are equipped with thermal compensation capabilities that correct for temperature changes from standard temperature (20℃). In any case, the CMM should be kept in a relatively clean environment and located in a space that is isolated from vibration.

A stationary bridge-type CMM is shown in Fig. 47.2, and the accuracy of this kind of CMM is usually better than that of a mobile articulated arm CMM. However, recent advances in the articulated arm area, have yielded significant advances in the capabilities of the articulated arm.

But for many operations, the accuracy of articulated arm CMMs is sufficient for a variety of processes. The advantage of articulated arm CMMs is that they generally have a larger work volume than bridge CMMs, and can reach areas that are not easy to access with typical CMMs. Thus, if quoted accuracies for articulated arm CMMs are sufficient for a particular application, it should be seriously considered as an alternative. Also, articulated arm CMMs are more portable. Typically, they can be set up for measurement quickly.

Figure 47.2 Bridge-type CMM

The size range of a CMM can span about four orders of magnitude with respect to part size. There are a variety of enormous CMMs that are used for measuring entire car bodies, the bodies of earth moving equipment, and even large aircraft elements (e.g., wings that are 10-

m long). There are other CMMs that measure parts that have features on the order of 1 mm. This capability can offer significant advances in micromanufacturing.

In the future, higher-speed measurements facilitated by linear drives and more advanced controls, in conjunction with thermal compensation, will be making further improvements in the CMM area. Other big changes may be expected in the software for CMMs as it becomes more user-friendly, and flexible. This will allow for much easier integration of the CMM into automated production facilities.

Words and Expressions

coordinate measuring machine (CMM) 三坐标测量机
digitize ['didʒitaiz] v. (将资料)数字化
configure [kən'figə] v. 使成形,为特定设备或用途而进行的设计、安排或设置
Cartesian coordinates 笛卡尔坐标,直角坐标
scanning ['skæniŋ] n. 扫描
linear motor 直线电动机
articulated [ɑː'tikjulitid] a. 铰接(的),枢接(的),有关节的
articulated arm CMM 关节臂测量机
servo ['sɔːvəu] n. 伺服系统,伺服机构
linear air bearing 直线空气轴承
linear scale 线性标度
turbine blade 涡轮机叶片
formidable ['fɔːmidəbl] a. 难对付的,难克服的,棘手的
controlled environment 受控环境
quoted ['kwəutid] a. 报出的,开出的,引证的
thermal compensation 热补偿
earth moving equipment 推土设备
micromanufacturing 微细制造,微细加工

linear drive 线性驱动

production facility 生产设备,生产设施

48. Computer Aided Process Planning

According to the *Tool & Manufacturing Engineers Handbook*, process planning is the systematic determination of the methods by which a product is to be manufactured economically and competitively. It essentially involves selection, calculation, and documentation. Processes, machines, tools, operations, and sequences must be selected. Such factors as feeds, speeds, tolerances, dimensions, and costs must be calculated. Finally, documents in the form of illustrated process sheets, operation sheets, and process routes must be prepared. Process planning is an intermediate stage between designing and manufacturing the product. But how well does it bridge design and manufacturing?

Most manufacturing engineers would agree that, if ten different planners were asked to develop a process plan for the same part, they would probably come up with ten different plans. Obviously, all these plans cannot reflect the most efficient manufacturing methods, and, in fact, there is no guarantee that any one of them will constitute the optimum method for manufacturing the part.

What may be even more disturbing is that a process plan developed for a part during a current manufacturing program may be quite different from the plan developed for the same or similar part during a previous manufacturing program and it may never be used again for the same or similar part. That represents a lot of wasted effort and produces a great many inconsistencies in routing, tooling, labor requirements, costing, and possibly even purchase requirements.

Of course, process plans should not necessarily remain static. As lot sizes change and new technology, equipment, and processes become available, the most effective way to manufacture a particular part also changes, and those changes should be reflected in current process plans released to the shop.

A planner must manage and retrieve a great deal of data and many documents, including established standards, machinability data, machine specifications, tooling inventories, stock availability, and existing process plans. This is primarily an information-handling job, and the computer is an ideal companion.

There is another advantage to using computers to help with process planning. Because the task involves many interrelated activities, determining the optimum plan requires many iterations. Since computers can readily perform vast numbers of comparisons, many more alternative plans can be explored than would be possible manually.

A third advantage in the use of computer-aided process planning is uniformity.

Several specific benefits can be expected from the adoption of compute-aided process planning techniques:

1. Reduced clerical effort in preparation of instructions.

2. Fewer calculation errors due to human error.

3. Fewer oversights in logic or instructions because of the prompting capability available with interactive computer programs.

3. Immediate access to up-to-date information from a central database.

4. Consistent information, because every planner accesses the same database.

5. Faster response to changes requested by engineers of other operating departments.

6. Automatic use of the latest revision of a part drawing.

7. More-detailed, more-uniform process-plan statements produced by word-processing techniques.

8. More-effective use of inventories of tools, gages, and fixtures and a concomitant reduction in the variety of those items.

9. Better communication with shop personnel because plans can be more specifically tailored to a particular task and presented in unambiguous, proven language.

10. Better information for production planning, including cutter life, forecasting, materials requirements planning, scheduling, and inventory control.

Most important for CIM, computer-aided process planning produces machine readable data instead of handwritten plans. Such data can readily be transferred to other systems within the CIM hierarchy for use in planning.

There are basically two approaches to computer-aided process planning: variant and generative.

In the variant approach, a set of standard process plans is established for all the parts families that have been identified through group technology. The standard plans are stored in computer memory and retrieved for new parts according to their family identification. Again, GT helps to place the new part in an appropriate family. The standard plan is then edited to suit the specific requirements of a particular job.

In the generative approach, an attempt is made to synthesize each individual plan using appropriate algorithms that define the various technological decisions that must be made in the course of manufacturing. In a truly generative process planning system, the sequence of operations, as well as all the manufacturing process parameters, would be automatically established without reference to prior plans. In its ultimate realization, such an approach would be universally applicable: present any plan to the system, and the computer produces the optimum process plan.

No such system exists, however. So called generative process-planning systems—and probably for the foreseeable future—are still specialized systems developed for a specific operation or a particular type of manufacturing process. The logic is based on a combination of past practice and basic technology.

Words and Expressions

process sheet 工艺过程卡,工艺卡
operation sheet 工序卡片
process route 工艺路线
retrieve [ri'tri:v] n. ;v. 检索
clerical ['klerikəl] a. 书写的,文书的,事物性的
prompt [prɔmpt] n. 提示
iteration [itə'reiʃən] n. 反复,重复,迭代
up-to-date ['ʌptə'deit] a. 现代化的,最新的,尖端的
concomitant [kən'kɔmitənt] a. 伴随的,随…而产生的
material requirement planning 物料需求计划(制造企业以在指定日期生产出指定产品为目标,确定生产所需的原料采购和部件装配的信息系统)
CIM 计算机集成制造
hierarchy ['haiərɑ:ki] n. 体系,系统,层次,分极结构
variant ['vɛəriənt] a. 不同的;n. 派生,衍生
variant approach 派生法
generative ['dʒenərətiv] a. 能生产的,创成的
generative approach 创成法
part family 零件族,零件组

49. Computer Numerical Control

Today, computer numerical control (CNC) machine tools are widely used in manufacturing enterprises. Computer numerical control is the automated control of machine tools by a computer and computer program.

The CNC machines still perform essentially the same functions as manually operated machine tools, but movements of the machine tool are controlled electronically rather than by hand. CNC machine tools can produce the same parts over and over again with very little variation. They can run day and night, week after week, without getting tired. These are obvious advantages over manually operated machine tools, which need a great deal of human interaction in order to do anything.

A CNC machine tool differs from a manually operated machine tool only in respect to the specialized components that make up the CNC system. The CNC system can be further divided into three subsystems: control, drive, and feedback. All of these subsystems must work together to form a complete CNC system.

1. Control System

The centerpiece of the CNC system is the control. Technically the control is called the machine control unit (MCU), but the most common names used in recent years are controller, control unit, or just plain control. This is the computer that stores and reads the program and tells the other components what to do.

2. Drive System

The drive system is comprised of screws and motors that will finally turn the part program into motion. The first component of the typical drive system is a high-precision lead screw called a ball screw. Eliminating backlash in a ball screw is very important for two reasons. First, high-precision positioning can not be achieved if the table is free to move slightly when it is supposed to be stationary. Second, material can be climb-cut safely if the backlash has been eliminated. Climb cutting is usually the most desirable method for machining on a CNC machine tool.

Drive motors are the second specialized component in the drive system. The turning of the motor will turn the ball screw to directly cause the machining table to move. Several types of electric motors are used on CNC control systems, and hydraulic motors are also occasionally used.

The simplest type of electric motor used in CNC positioning systems is the stepper motor (sometimes called a stepping motor). A stepper motor rotates a fixed number of degrees when it receives an electrical pulse and then stops until another pulse is received. The stepping characteristic makes stepper motors easy to control.

It is more common to use servomotors in CNC systems today. Servomotors operate in a smooth, continuous motion-not like the discrete movements of the stepper motors. This smooth motion leads to highly desirable machining characteristics, but they are also difficult to control. Specialized hardware controls and feedback systems are needed to control and drive these motors. Alternating current (AC) servomotors are currently the standard choice for industrial CNC machine tools.

3. Feedback System

The function of a feedback system is to provide the control with information about the status of the motion control system, which is described in Figure 49.1.

The control can compare the desired condition to the actual

condition and make corrections. The most obvious information to be fed back to the control on a CNC machine tool is the position of the table and the velocity of the motors. Other information may also be fed back that is not directly related to motion control, such as the temperature of the motor and the load on the spindle—this information protects the machine from damage.

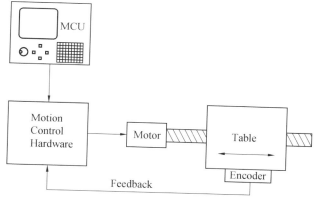

Figure 49.1　Typical motion control system of a CNC machine tool

There are two main types of control systems: open-loop and closed-loop. An open-loop system does not have any device to determine if the instructions were carried out. For example, in an open-loop system, the control could give instructions to turn the motor 10 revolutions. However, no information can come back to the control to tell it if it actually turned. All the control knows is that it delivered the instructions. Open-loop control is not used for critical systems, but it is a good choice for inexpensive motion control systems in which accuracy and reliability are not critical.

Closed-loop feedback uses external sensors to verify that certain conditions have been met. Of course, positioning and velocity feedback is of primary importance to an accurate CNC system. Feedback is the only way to ensure that the machine is behaving the way the control intended it to behave.

Words and Expressions

subsystem ['sʌbsistim] n. 子系统
centerpiece ['sentəpi:s] n. 主要特征
drive system 驱动系统,传动系统
backlash ['bæklæʃ] n. 反向间隙,回程误差
climb cutting 顺切,顺铣
drive motor 驱动电动机
ball screw 滚珠丝杠副
hydraulic motor 液压马达
stepper motor 步进电机
servomotor ['sə:vəu,məutə] 伺服电动机
feedback ['fi:dbæk] n. 反馈
alternating current 交流,交流电
open loop 开环
closed loop 闭环
encoder [in'kəudə] n. 编码器

50. AI: Promises Start to Pay Off

Is artificial intelligence (AI) for real? How much promise does it really have in industrial automation? The answers vary widely, depending on whom you ask. Although many people in manufacturing think of AI as only a distant possibility, the fact is that segments of this fast-developing technology have much to offer right now. Having the greatest impact today are knowledge-based expert systems designed to complement conventional computer-integrated manufacturing. These AI systems use computers in ways that are markedly different from those of conventional data processing, harnessing the qualities of human reasoning to solve problems.

Expert systems are already being used successfully in industry. Equipment diagnostic applications, for instance, are reducing equipment downtime and maintenance costs by assisting technicians in finding hardware problems. In the area of alarm analysis, expert systems are helping to monitor processes in real time, interpreting alarm conditions and determining corrective actions. By decreasing reaction time, the losses incurred due to a mishandled alarm can be reduced or eliminated.

On the production floor, expert systems are being used as operator aids, assisting workers to operate machinery. These systems not only provide a constant level of expertise at key operator stations, but also produce a variety of status, production, and other information reports to assist plant management.

Before a plant is built, process planning expert systems can be used to establish plant floor management and to simulate and diagnose processes. The cost of such systems can be offset easily by the reduction in planning time and elimination of scrap and less-than-nominal products. Once built, the plant can rely on knowledge-based scheduling systems to manage the variables involved in production process scheduling.

Systems of this type are also being developed to improve the efficiency of the manufacturing process itself. These tools monitor and coordinate flexible manufacturing systems on a real-time basis, adapting to changes that occur on the plant floor.

Along with expert systems—and, in many cases, in conjunction with them—AI is becoming more commonplace in the form of man-machine interfaces, robotics, and vision systems. Man-machine interfaces have been improved through the use of color displays, high-resolution graphics, and touch screens. Speech synthesis and recognition greatly facilitate such communication and are being used more and more in industry. In the control room, for example, voice output systems are replacing bells and blinking lights. On the factory or warehouse floor, speech recognition systems are helping speed inventory control and quality inspection by eliminating the need to manually enter data.

Another man-machine interface area where AI is being used today is natural language. Natural language interface processing allows users to interface to computers in English-like phrases instead of using rigid, meaningless computer commands.

Perhaps the most visible—and in some ways least developed—applications for AI are in the area of robotics and vision systems. State-of-the-art robotic systems are rather simplistic compared to their human counterparts. Nevertheless, robot development has moved ahead in many areas—for example, welding, material handling, parts positioning, assembly, and spray painting. Growth in machine vision has been spurred by industry's increased focus on quality. Until recently, most inspection tasks were performed manually. Now, the trend is toward using vision, especially in the automotive and electronics industries. Today, vision systems typically are used by themselves on factory lines, where they inspect, identify, and measure parts and assemblies, rather than being used in conjunction with robots.

The theory behind expert systems is to capture the knowledge of human experts in specialized areas or domains and apply this knowledge to problem solving. Today, these systems are typically developed through the assistance of knowledge engineers who gather the information through an interviewing process representing the rules-of-thumb models and facts in computer programs. For some applications, expert system "shells" can be used to speed up the development process. Prototypes are created first, with the knowledge base expanded and modified until the system reaches expert performance.

It's important to note that expert systems are tools, not replacements for human labor. An expert system can, in many cases, free personnel to extend their expertise to other areas. When key personnel retire, transfer, or quit, expert systems can ease the transition and serve as a training tool for new personnel.

But there are a number of barriers to developing knowledge-based systems. One is a psychological resistance on the part of company "experts". There is often the fear that, upon exposing their knowledge, they will appear less expert in their specialty. Along with this is the worry of being replaced once a computer can handle the job. Other pitfalls include gathering an insufficient amount of data to effectively address all the problem-solving requirements of an application, and leaving the actual users out of the development process, thus creating resistance to use of the system.

Words and Expressions

for real 真实地,确定地,现实确实如此地
segment ['seɡmənt] n. 部分,片段
data processing 数据处理
harness ['hɑːnis] v. 利用,治理
reasoning ['riːzəniŋ] n. 推理,推论; a. 理性的,推理
downtime ['dauntaim] n. 停机时间,发生故障时间,修理时间
incur [in'kəː] v. 招致,承担,遭受

commonplace ['kɔmənpleis] a. 平凡的,平常的;n. 平淡的事
color display 彩色显示器
high-resolution graphics 高分辨率图形
touch screen 触摸式屏幕
speech synthesis and recognition 语音合成与识别
blinking ['bliŋkiŋ] a. 闪烁的
warehouse ['wɛəhaus] n. 库房,储存室
inventory control 库存管理
natural language 自然语言(人类写或说的语言)
simplistic [sim'plistik] a. 过分简单的
spray painting 喷漆
spur [spə:] v. 激励,鞭策,鼓励
in conjunction with 同……一道,连同
knowledge engineer 知识工程师,建立专家系统的编程人员
psychological [saikə'lɔdʒikəl] a. 心理的
on the part of 在…方面
pitfall ['pitʃɔ:l] n. 缺陷,易犯的错误

51. Online Design

As design cycle time speeds ever faster, the method of design itself continually changes to keep up with the pace. Computer-based technologies have greatly influenced the way design engineers work. But more is on the way, engineers and executives are urged to take advantage of three technologies in particular.

The first technological innovation—now well entrenched—was the use of high-powered personal computers. With PCs, engineers had access to high-speed applications of computer-aided-design software right at their own desks. Personal computers took the place of rulers and pencils.

The second innovation is the advancing capability of PCs to function as supercomputers. By taking advantage of this technology, engineers untrained in a mathematical application such as finite element analysis can run an FEA software program that performs calculations automatically and will shave weeks off the design process. Previously, the designs had been sent to an FEA professional for analysis, with a turnaround time that was weeks long.

A third change, which came about after the other developments laid the groundwork, is the emerging use of the Internet by engineers, who are increasingly designing in collaboration with partners in far-flung locations.

The Internet is seeing other uses in the engineering industry as well. Some vendors host software on their own servers and, for a fee, offer engineers access to programs via the Internet. This eliminates the need for engineers to pay for the software for longer than they need it. They do not have to download it to their own computers.

Software vendors are functioning as what are called application server providers. That is, the vendor is providing the application, which resides on its server. The engineers are merely accessing the application. They don't own it or use it under an annually renewed license. Instead, they subscribe to the application, often on a month-by-month or per-use basis.

At the Web site, users can choose to license the software for a set amount of time. After subscribing, they go to the site each time they want to use the software and click on an icon. That boots up the software and the user's individual home page, also maintained on the vendor's server.

The ability to buy monthly term licenses cuts costs considerably for engineering companies that may plan to use the program for a short time. Also, the application doesn't reside on the user's personal computer. In fact, while the software performs its analysis, the engineer can minimize the screen and open an Excel spreadsheet or a word processing program to do other work. The complicated analysis doesn't slow down other applications or prevent them from running.

New analysis capabilities are speeding the pace of design, which, in turn, nudges engineers toward the need for networked capabilities and access to online applications. Engineers no longer need advanced training in finite element analysis, for example. Now, the computer program does the analysis for them, allowing them to analyze while they design. Simulation software is also being used in all areas of manufacturing and consumer design.

Another relatively recent advance, the capability to create the design in a mathematical representation called solid modeling, also is pushing the advance of online and networked applications. A computerized solid model provides a crisp, clear view of the product design, suitable for passing to manufacturing engineers and marketers within the engineering company.

Many companies will find themselves taking advantage of the Internet or of an intranet, which, when coupled with a firewall or with encryption methods to ensure confidentiality, can act as a private network for company employees to communicate back and forth, or to communicate with vendors, suppliers, and manufacturers.

Engineering companies can use the Web-based technology quickly because they don't need to buy software or, often, even hardware. Companies usually have Web-enabled personal computers.

The nature of the Web also means that engineers can communicate with each other in real time on the same project, in a cooperative process often called "collaborative engineering." A collaborative engineering environment can take place entirely over the

Web.

These engineers can work together in real time from distant locations to come up with the best design for assembly. Then, after they come up with it, they can say, "Let's go and sell it." Not only does this engineering environment pull in design engineers, but it also pulls in product managers, manufacturing engineers, marketers, and senior management. They all have access to the system.

The engineers usually have to sell the design to the senior management and this Web capability allows them to do that very easily. The managers can see the design in real time and if they ask a what-if question, like what would happen if you changed this geometry, they can run the simulation and see the results of what would happen right on the spot.

Manufacturing engineers have access to the same online design in order to study up-close how to manufacture the product. They could also ask mechanical engineers questions about the design and could run simulations, if that should be necessary to get a question answered.

The growing Internet bandwidth and a single voice and data fiber will soon let engineers talk and see each other in video images that will appear on one corner of the computer screen. That will enable them to guide each other through design images on the screen, as they speak.

Words and Expressions

online [ˈɔnlain] a. 联机的,在线的
on the way 在途中
entrench [inˈtrentʃ] v. 牢固树立,确立,确定
high-powered a. 大功率的,高性能的

finite element analysis (FEA) 有限元分析
turnaround 转变,转向,突然好转
turnaround time 解题周期,周转时间
groundwork ['graundwə:k] n. 基础,根本,基本原理
emerging [i'mə:dʒiŋ] a. 新兴的,新出现的
far-flung 分布广的,范围广的,遥远的,漫长的
server 服务器
access into 使用,进入
application 应用程序,应用软件
application server 应用服务器
reside [rizaid] v. 驻留,居住,归于
Web site 万维网站,万维站点
click on 在…上单击
icon ['aikɔn] n. 图标,肖像
boot up 引导,使系统开始工作
home page 主页
spreadsheet 电子制表软件,电子数据表,电子表格
nudge [nʌdʒ] n. 轻推,接近,推进,刺激
solid modeling 实体造型
crisp [krisp] a. 清新的,明快的
marketer ['mɑ:kitə] n. 市场上的买卖双方,销售经理
Intranet [intrə'net] n. 企业内部互联网
firewall 防火墙
encryption [in'kripʃən] n. 加密,编密码
come up with 赶上,提出,拿出
pull in 获得,吸引
what-if 假设分析,作假定推测
on the spot 当场,处于负责地位
up-close 在很近距离内
bandwidth ['bændwidθ] n. 带宽,频带宽度

fiber 纤维,光纤

52. Flexible Manufacturing Systems

Flexibility is an important characteristic in the modern manufacturing. It means that a manufacturing system is versatile and adaptable, while also capable of handling relatively high production runs. A flexible manufacturing system (FMS) integrates all major elements of manufacturing into a highly automated system. First utilized in the late 1960s, FMS consists of a number of manufacturing cells, each containing an industrial robot (serving several CNC machines) and an automated material-handling system, all interfaced with a central computer. Different computer instructions for the manufacturing process can be downloaded for each successive part passing through the manufacturing cell.

This system is highly automated and is capable of optimizing each step of the total manufacturing operation. These steps may involve one or more processes and operations (such as machining, grinding, heat treating, and finishing), as well as handling of raw materials, inspection, and assembly. The most common applications of FMS to date have been in machining and assembly operations. A variety of FMS technology is available from machine tool manufacturers.

Flexible manufacturing systems represent the highest level of efficiency, sophistication, and productivity that has been achieved in manufacturing plants. The flexibility of FMS is such that it can handle a variety of part configurations and produce them in any order. FMS can be regarded as a system which combines the benefits of two other systems: (a) the highly productive but inflexible transfer lines, and (b) job-shop production, which can produce large product variety on stand-alone machines but is inefficient.

In FMS, the time required for changeover to a different part is very short. The quick response to product and market-demand variations is a major attribute of FMS.

The basic elements of a flexible manufacturing system are (a) manufacturing cells, (b) automated handling and transport of materials and parts, and (c) control systems. The manufacturing cells are arranged to yield the greatest efficiency in production, with an orderly flow of materials, parts, and products through the system.

The types of machines in manufacturing cells depend on the type of production. For machining operations, they usually consist of a variety of three- to five-axis machining centers, CNC lathes, milling machines, drill presses, and grinders. Also included is various other equipment, such as that for automated inspection (including coordinate measuring machines), assembly, and cleaning.

Because of the flexibility of FMS, material-handling, storage, and retrieval systems are very important. Material handling is controlled by a central computer and performed by automated guided vehicles, conveyors, and various transfer mechanisms. The system is capable of transporting raw materials, blanks, and parts in various stages of completion to any machine (in random order) and at any time: Prismatic parts are usually moved on specially designed pallets. Parts having rotational symmetry (such as those for turning operations) are usually moved by mechanical devices and robots.

The computer control system of FMS is its brains and includes various software and hardware. This sub-system controls the machinery and equipment in manufacturing cells and the transporting of raw materials, blanks, and parts in various stages of completion from machine to machine. It also stores data and provides communication terminals that display the data visually.

FMS installations are very capital-intensive. Consequently, a thorough cost-benefit analysis must be conducted before a final decision is made. This analysis should include such factors as the cost of capital, of energy, of materials, and of labor, the expected markets for the products to be manufactured, and any anticipated fluctuations in market demand and product type. An additional factor is the time and effort required for installing and debugging the system. Although FMS requires few, if any, machine operators, the personnel in charge of the total operation must be trained and highly skilled. These personnel include manufacturing engineers, computer programmers, and maintenance engineers. The most effective FMS applications have been in medium-volume batch production. In contrast, high-volume, low-variety parts production is best obtained from transfer machines (dedicated equipment). Finally, low-volume, high-variety parts production can best be done on conventional standard machinery (with or without NC) or by machining centers.

Words and Expressions

versatile ['və:sətail] *a.* 通用的,多用途的
adaptable [ə'dæptəbl] *a.* 能适应的,适应性强的,可修改的
flexible manufacturing system (FMS) 柔性制造系统
manufacturing cell 制造单元,加工单元
material handling system 物料传送系统,物料搬运系统
central computer 中央计算机
download [daun'ləud] *v.* 传送,下载
successive [sək'sesiv] *a.* 逐次的,相继的,连续的
manufacturing plant 制造厂
configuration [kən‚figju'reiʃən] *n.* 构造,结构,配置,外形
inflexible transfer line 刚性自动生产线

job-shop production 单件小批生产
changeover ['tʃeindʒ'əuvə] n. 变换,转换
machining center 加工中心
automated guided vehicle 自动导向输送车
transfer mechanism 输送机构,机械手
prismatic [priz'mætik] a. 棱形的,棱柱形的
rotational symmetry 轴向对称,旋转对称
turning operation 车削操作,车削加工
retrieval [ri'triːvəl] n. 检索,查找
random order 任意顺序,随机次序
communication terminal 通信终端,系统或网络中可以发送或接收数据的终端
cost of capital 资金成本,资本成本
maintenance engineer 技术维护工程师
batch production 成批生产,分批生产,批量生产

53. Computer-Integrated Manufacturing

Beginning with computer graphics and computer-aided modeling, design, and manufacturing, the use of computers has been extended to computer-integrated manufacturing (CIM) in which software and hardware are integrated from product concept through product distribution in the marketplace. Computer-integrated manufacturing is particularly effective because of its capability for making possible:

1. Responsiveness to rapid changes in market demand and product modifications.

2. Better use of materials, machinery, and personnel, and reduction in inventory.

3. Better control of production and management of the total manufacturing operation.

4. The manufacture of high-quality products at low cost.

The following is an outline of the major applications of computers in manufacturing:

1. **Computer numerical control** (**CNC**). Numerical control (CNC) machine tools were developed before the availability of inexpensive computing power and therefore had an electronic control but not a genuine computerized control. Computer numerical control (CNC) is a method of controlling the movements of machine components by a computer and computer program. The terms NC and CNC are used synonymously today, although actual NC machines are becoming rare in most shops. Regardless of the control type, the code that is used to produce the parts is known as NC code-not CNC code.

2. **Adaptive control** (**AC**). The parameters in a manufacturing process are adjusted automatically to optimize production rate and product quality, and to minimize cost. Parameters such as forces, temperatures, surface roughness, and dimensions of the part are monitored constantly; if they move outside the acceptable range, the system adjusts the process variables until the parameters again fall within the specified range.

3. **Industrial robots**. Since 1960s, industrial robots have been replacing humans in operations that are repetitive, dangerous, and boring, thus reducing the possibility of human error, decreasing variability in product quality, and improving productivity. Robots with sensory perception capabilities have been developed (intelligent robots), with movements that simulate those of humans.

4. **Automated handling of materials**. Computers have made possible highly efficient handling of materials and components in various stages of manufacture, such as when being moved from storage to machines, from machine to machine, and at points of inspection, inventory, and shipment.

5. Automated and robotic assembly systems. These systems mainly have replaced costly assembly by human operators, although humans still have to perform some of these operations. Products are now designed or redesigned so that they can be assembled more easily and faster by machines.

6. Computer-aided process planning (CAPP). This system is capable of improving productivity by optimizing process plans, reducing planning costs, and improving the consistency of product quality and reliability. Functions such as cost estimating and monitoring of work standards (time required to perform a certain operation) are also incorporated into the system.

7. Group technology (GT). The concept of group technology is that parts can be grouped and produced by classifying them into families, according to similarities in design and similarities in the manufacturing processes employed to produce the part. In this way, part designs and process plans can be standardized and families of similar parts can be produced efficiently and economically.

8. Just-in-time production (JIT). The principle of JIT is that supplies of raw materials, parts, and components are delivered to the manufacturer just in time to be used, parts and components are produced just in time to be made into subassemblies and assemblies, and products are finished just in time to be delivered to the customer. As a result, inventory costs are low, part defects are detected right away, productivity is increased, and high-quality products are made at low cost.

9. Expert systems (ES). These systems basically are complex computer programs; they have the capability to perform various tasks and solve difficult real-life problems much as human experts would.

10. Artificial intelligence (AI). This important field involves the use of machines and computers to replace human intelligence.

Computer-controlled systems are now capable of learning from experience and of making decisions that optimize operations and minimize costs. Artificial neural networks (ANN), which are designed to simulate the thought processes of the human brain, have the capability of modeling and simulating production facilities, monitoring and controlling manufacturing processes, diagnosing problems in machine performance, conducting financial planning, and managing a company's manufacturing strategy.

In view of these continuing advances and their potential, some experts have envisioned the factory of the future. Although highly controversial, and viewed as unrealistic by some, this is a system in which production will take place with little or no direct human intervention. Ultimately, the human role is expected to be confined to the supervision, maintenance, and upgrading of machines, computers, and software.

Words and Expressions

computer graphics 计算机制图,计算机图形学,计算机制图学
modeling ['mɔdliŋ] *n.* 建模,造型
computer-integrated manufacturing (CIM) 计算机集成制造,计算机集成制造技术
inventory ['invəntri] *n.* 存货,库存
manufacturing operation 制造过程,生产制造过程
outline ['əutlain] *n.* 要点,概要,大纲,轮廓
adaptive control 自适应控制
sensory perception 感官知觉
intelligent robot 智能机器人
redesign [ˌriːdi'zain] *v.* 重新设计; *n.* 重新设计,新设计
work standard 作业标准

group technology 成组技术
classify ['klæsifai] v. 分类,分等
manufacturing process 生产过程
just in time production 准时生产
expert system 专家系统
artificial intelligence (AI) 人工智能
artificial neural network (ANN) 人工神经网络
thought process 思维过程
production facilities 生产设备
intervention [ˌintə'venʃən] n. 介入,干涉,干预

54. Mechanical Engineering in the Information Age

In the early 1980s, engineers thought that massive research would be needed to speed up product development. As it turns out, less research is actually needed because shortened product development cycles encourage engineers to use available technology. Developing a revolutionary technology for use in a new product is risky and prone to failure. Taking short steps is a safer and usually more successful approach to product development.

Shorter product development cycles are also beneficial in an engineering world in which both capital and labor are global. People who can design and manufacture various products can be found anywhere in the world, but containing a new idea is hard. Geographic distance is no longer a barrier to others finding out about your development six months into the process. If you've got a short development cycle, the situation is not catastrophic as long as you maintain your lead. But if you're in the midst of a six-year development process and a competitor gets wind of your work, the project could be in more serious trouble.

The idea that engineers need to create a new design to solve every problem is quickly becoming obsolete. The first step in the modern design process is to browse the Internet or other information systems to see if someone else has already designed a transmission, or a heat exchanger that is close to what you need. Through these information systems, you may discover that someone already has manufacturing drawings, numerical control programs, and everything else required to manufacture your product. Engineers can then focus their professional competence on unsolved problems.

In tackling such problems, the availability of high-powered personal computers and access to the information highway dramatically enhance the capability of the engineering team and its productivity. These information age tools can give the team access to massive databases of material properties, standards, technologies, and successful designs. Such pretested designs can be downloaded for direct use or quickly modified to meet specific needs. Remote manufacturing, in which product instructions are sent out over a network, is also possible. You could end up with a virtual company where you don't have to see any hardware. When the product is completed, you can direct the manufacturer to drop-ship it to your customer. Periodic visit to the customer can be made to ensure that the product you designed is working according to the specifications. Although all of these developments won't apply equally to every company, the potential is there.

Custom design used to be left to small companies. Big companies sneered at it they hated the idea of dealing with niche markets or small-volume custom solutions. "Here is my product," one of the big companies would say. "This is the best we can make it you ought to like it. If you don't, there's a smaller company down the street that will work on your problem."

Today, nearly every market is a niche market, because customers are selective. If you ignore the potential for tailoring your product to specific customers' needs, you will lose the major part of your market share perhaps all of it. Since these niche markets are transient, your company needs to be in a position to respond to them quickly.

The emergence of niche markets and design on demand has altered the way engineers conduct research. Today, research is commonly directed toward solving particular problems. Although this situation is probably temporary, much uncommitted technology, developed at government expense or written off by major corporations, is available today at very low cost. Following modest modifications, such technology can often be used directly in product development, which allows many organizations to avoid the expense of an extensive research effort. Once the technology is free of major obstacles, the research effort can focus on overcoming the barriers to commercialization rather than on pursuing new and interesting, but undefined, alternatives.

When viewed in this perspective, engineering research must focus primarily on removing the barriers to rapid commercialization of known technologies. Much of this effort must address quality and reliability concerns, which are foremost in the minds of today's consumers. Clearly, a reputation for poor quality is synonymous with bad business. Everything possible, including thorough inspection at the end of the manufacturing line and automatic replacement of defective products, must be done to assure that the customer receives a properly functioning product.

Research has to focus on the cost benefit of factors such as reliability. As reliability increases, manufacturing costs and the final cost of the system will decrease. Having 30 percent junk at the end of a production line not only costs a fortune but also creates an opportunity for a competitor to take your idea and sell it to your customers.

Central to the process of improving reliability and lowering costs is the intensive and widespread use of design software, which allows engineers to speed up every stage of the design process. Shortening each stage, however, may not sufficiently reduce the time required for the entire process. Therefore, attention must also be devoted to concurrent engineering software with shared databases that can be accessed by all members of the design team.

As we move more fully into the Information Age, success will require that the engineer possess some unique knowledge of and experience in both the development and the management of technology. Success will require broad knowledge and skills as well as expertise in some key technologies and disciplines; it will also require a keen awareness of the social and economic factors at work in the marketplace. Increasingly, in the future, routine problems will not justify heavy engineering expenditures, and engineers will be expected to work cooperatively in solving more challenging, more demanding problems in substantially less time. We have begun a new phase in the practice of engineering. It offers great promise and excitement as more and more problem-solving capability is placed in the hands of the computerized and wired engineer. To assure success, the capability of our tools and the unquenched thirst for better products and systems must be matched by the joy of creation that marks all great engineering endeavors. Mechanical engineering is a great profession, and it will become even greater as we make the most of the opportunities offered by the Information Age.

Words and Expressions

turn out 结果是,原来是,证明是
prone [prəun] *a.* 有…倾向,易于…的
geographic [dʒiə'græfik] *a.* 地理上的,地区性的
get wind of... 获得…线索,听到…风声
browse [brauz] *v.* 浏览,翻阅
transmission [trænz'miʃən] *n.* 传动装置,变速箱
tackle ['tækl] *v.* (着手)处理,从事,对付,解决
information highway 信息高速公路
information age 信息时代
pretest ['priːtest] *n.* ;*v.* 事先试验,预先测试
end up with... 以…为结束,最后得出
virtual company 虚拟公司
niche market 瞄准机会的市场(指专门瞄准机会,做因市场不大而别人不做的产品,从而获得丰厚的利润)
tailor... to... 使…适合[满足]…(要求,需要,条件等)
transient ['trænziənt] *a.* 短暂的,瞬变的
confer [kən'fəː] *v.* 授予,给予,使具有(性能)
uncommitted ['ʌnkə'mitid] *a.* 自由的,不受约束的,不负义务的
write off 抹去,注销,取销
undefined ['ʌndi'faind] *a.* 未规定的,不明确的,模糊的
junk [dʒʌŋk] *n.* 废物,废品;*v.* 把…(当作废物)丢弃
be synonymous with... 和…同义,和…的意义是一样的
expertise [ekspəˈtiːz] *n.* 专门知识,专长,专家鉴定
unquenchable [ʌnˈkwentʃəbl] *a.* 不能熄灭的,不能遏制的
endeavor [inˈdevə] *n.* ;*v.* 努力,尽力,事业,活动

55. Mechatronics

The success of industries in manufacturing and selling goods in a world market increasingly depends upon an ability to integrate electronics and computing technologies into a wide range of primarily mechanical products and processes. The performance of many current products —cars, washing machines, robots or machine tools—and their manufacture depend on the capacity of industry to exploit developments in technology and to introduce them at the design stage into both products and manufacturing processes. The result is systems which are cheaper, simpler, more reliable and with a greater flexibility of operation than their predecessors. In this highly competitive situation, the old divisions between electronic and mechanical engineering are increasingly being replaced by the integrated and interdisciplinary approach to engineering design referred to as mechatronics.

In a highly competitive environment, only those new products and processes in which an effective combination of electronics and mechanical engineering has been achieved are likely to be successful. In general, the most likely cause of a failure to achieve this objective is an inhibition on the application of electronics. In most innovative products and processes the mechanical hardware is that which first seizes the imagination, but the best realization usually depends on a consideration of the necessary electronics, control engineering and computing from the earliest stages of the design process. The integration across traditional boundaries lies at the heart of a mechatronic approach to engineering design and is the key to understanding the developments that are taking place.

To be successful, a mechatronic approach needs to be established from the very earliest stages of the conceptual design process, where options can be kept open before the form of embodiment is determined. In this way the design engineer, and especially the mechanical design engineer, can avoid going too soon down familiar and perhaps less productive paths.

Where full attention has been given to market trends, the adoption of an integrated mechatronic approach to design has led to a revival in areas such as high speed textile equipment, metrology and measurement systems, and special purpose equipment such as that required for the automatic testing of integrated circuits. In most cases the revival or new growth is brought about by the enhancement of process capability achieved by the integration of electronics, often in the form of an embedded microprocessor, with the basic mechanical system.

This demand for increased flexibility in the manufacturing process has led to the development of the concept of flexible manufacturing systems (FMSs) in which a number of elements such as computer numerically controlled machine tools, robots and automatically guided vehicles (AGVs) are linked together for the manufacture of a group of products. Communication between the individual elements of the system is achieved by means of local area networks (LANs).

Within products the diversity and opportunity offered by a mechatronic approach to engineering design is to date largely unrealized. End user products are substantial revenue earners and it is possible here to distinguish between existing products offering enhanced capabilities and completely new product areas which would not have existed without a mechatronic design approach having been adopted from the outset.

In the first category the following are illustrative from many examples:

Automotive engines and transmissions Engine and driveline management systems leading to reduced emissions, improved fuel economy, protection against driver misuse by, for example, prohibiting excessive fuel flow at low speeds, and selectable gear characteristics.

Power tools Modern power tools such as drills offer a range of features including speed and torque control, reversing drives and controlled acceleration.

Examples in the second category include the following:

Modular robots Conventional industrial robots are often limited in their operation by their geometry. By providing a range of structural components and actuators together with a central controller a modular robot system has been made available, allowing users to assemble robot structures directly suited to their needs.

Video and compact disc players Video and compact disc systems involve complex laser tracking systems to read the digitally encoded signal carried by the disc. This control is achieved by means of a microprocessor based system which also provides features such as multiple track selection, scanning and preview.

A common factor in consumer mechatronics as exemplified by the above is the continuous improvement in capability achieved against a constant or reducing real cost to the end user. The capability of a mechatronic system, based as it often is on inexpensive components or modules, also provides a means to execute bespoke solutions to special problems.

In engineering design, a mechatronics approach represents and requires the integration of a wide range of material and information aimed at providing systems which are more flexible and of higher performance than their predecessors. Thus, for full benefit and effect, mechatronics must be a feature of both the conceptual and embodiment phases of the design process.

In manufacturing, users are demanding a much higher degree of control of both the overall process and its components. This requires a knowledge both of the capabilities of these components and of the means by which they are integrated within the complete system.

Words and Expressions

exploit [iks′plɔit] v. 开发,利用,使用,发挥;n. 功绩,成就
innovative [′inəuveitiv] a. 革新的,创新的,富有革新精神的
seize [si:z] v. 抓,捉,了解,利用
lie at the heart of 是…的核心(精华)所在
embodiment [im′bɔdimənt] n. 具体化,具体表现,具体装置
revival [ri′vaivəl] n. 复兴,恢复,再生
metrology [mi′trɔlədʒi] n. 计量学,测量学,计量制
bring about 引起,产生,造成,促使,导致,完成
enhancement [in′hɑ:nsmənt] n. 增强,提高,放大
automatically guided vehicle 自动导引小车,自动制导车辆
local area network 局域网,本地网
diversity [dai′və:siti] n. 不同,异样性,多种多样性
to date 至今,到目前为止,截止当天
end user 终端用户,最终用户
revenue [′revinju:] n. 收入,收益,进款
outset [′autset] n. 开端,开始,最初
illustrative [′iləstritiv] a. 说明的,解说的,例证的
drive line 动力传动系统,传动轴装置
emission [i′miʃən] n. 发射,放出,排出物
misuse [′mis′ju:z] v. ;n. 错用,误用,滥用
power tool 电动工具
modular [′mɔdjulə] a. 制成标准组件的,预制的,组合的

track [træk] n. 轨道,磁道,导向装置;v. 跟踪,沿轨道行驶
scan [skæn] v. 检查,扫描,浏览
preview ['priː'vjuː] n. ;v. 预览,事先查看,预演
module ['mɔdjuːl] n. 模数,模件,组件,可互换标准件
bespoke [bi'spəuk] a. 专做订货的;n. 预订的货
execute ['eksikjuːt] v. 实行,完成,实现,实施
exemplify [ig'zemplifai] v. 例证,举例说明,作为…的例子

6 EDUCATION

56. Manufacturing Engineering Education

A manufacturing engineering program accredited

Miami University (Oxford, Ohio) has recently received accreditation for its Manufacturing Engineering Program by the Engineering Accreditation Commission of the Accreditation Board for Engineering and Technology (ABET). Miami's program becomes one of only ten accredited such programs in the entire nation. In addition, Miami is the only university in the five-state region of Ohio, Indiana, Kentucky, Pennsylvania, and West Virginia to have an ABET-accredited manufacturing engineering program.

The objective of Miami's program is to produce engineering graduates who can design, analyze, and apply manufacturing methods and processes so that quality products can be produced at a competitive cost. The broad-based curriculum contains elements of electrical, mechanical and industrial engineering with a focus on manufacturing processes. It emphasizes problem solving and design, and integrates concepts with laboratory experience in computer-aided design, computer-controlled manufacturing systems, statistical process control, microprocessors, electronic instrumentation and materials testing. In addition to the rigorous engineering core, the curriculum includes courses in humanities and social sciences and enhances the development of communication skills and teamwork.

Graduates of Miami's program typically work as manufacturing engineers in areas such as product and process design, quality control, computer-aided manufacturing and plant facilities engineering. Starting salaries for Miami's graduates in 1990 averaged more than $30,000 and more than 90% of the graduates had accepted job offers before graduation.

MIT launches manufacturing program with industry

Nine of the nation's leading manufacturing corporations have joined with Massachusetts Institute of Technology to launch an educational and research program aimed at helping the U.S. recapture world leadership in manufacturing. Participating companies are Alcoa, Boeing Co., Digital Equipment Corp., Hewlett-Packard, Johnson & Johnson, Kodak, Motorola, Polaroid, and United Technologies Corp. Three additional firms will participate and will include representation from the auto industry.

As part of the initiative, MIT has established a novel two-year master's program, called Leaders for Manufacturing, in which students receive two degrees—one from the Sloan School of Management and one from a department in the School of Engineering. The first class of students started in June 1988 and the 1989 class is currently being recruited. Graduates of the program will have both the technical and managerial skills required to lead sophisticated manufacturing projects and operations that take advantage of the latest technologies.

At the same time, MIT has embarked on a cooperative program of research with industry to identify the critical issues and knowledge base needed for manufacturing leadership. Principles that will be taught, and ultimately practiced in industry, will be established through multidisciplinary investigations of all phases of the manufacturing process—design, development, production, marketing, delivery, and service. Eventually, 10% to 15% of the faculty in the schools of engineering and management will be directly involved in the program.

The Leaders in Manufacturing Program will investigate the various components of complex manufacturing operations based on systems approach, looking at the whole manufacturing process. Students in the program will be required to spend up to six months in an industrial manufacturing environment under close faculty and industrial supervision. In order to guarantee the practical focus and usefulness of the initiative, the industrial partners also are actively participating in both the educational developments and the research projects. Senior executives from each firm have committed substantial time to help senior MIT faculty guide the overall course of the program.

In addition, key company personnel are working side-by-side with MIT faculty to develop specific research projects and new subjects. MIT has named at least one professor to be the liaison with each participating company with the creation of a series of chaired professorships. The partner firms and MIT have committed substantial resources (people, facilities, and funds) to make a difference in the world of manufacturing, research, and education.

Furthermore, the industrial partners are providing opportunities for teams of students and faculty to work with industry manufacturing experts on real problems encountered in their own manufacturing facilities. A complementary program of on-campus research has been initiated in collaboration with the industrial firms. Over the next five years several hundred students and faculty are expected to participate in the program.

A manufacturing systems engineering program

University of Pittsburgh has established a new graduate degree program for working engineers that will teach them how to design and control the manufacturing process in all its cycles from product conceptualization to retirement. The program was established with the help of a $1.4 million grant from Westinghouse Electric Corp. and the support of Alcoa, PPG Industries, General Motors, and others.

Called the Manufacturing Systems Engineering Program, or MSEP, it is aimed primarily at practicing engineers with two to three years of work experience. Under the direction of Dr. John H. Manley, the School of Engineers is wrapping up its first full year of MSEP instruction with 38 students already enrolled. The Master of Science degree involves one year of study, or 36 credits, including a six credit-hour, on-the-job internship. The requirements for this degree are heavily engineering-oriented and are intended to be consistent in academic content and rigor with the School of Engineering's other master's programs. Multiple disciplines are integrated into the MSEP. These include not only the traditional engineering and materials science disciplines, but software engineering, computer engineering, human factors, and engineering management.

In addition to traditional engineering subjects, elective courses that emphasize international competitiveness have been developed for MSEP by faculty in the Joseph M. Katz Graduate School of Business. Material for other course modules has been developed jointly by School of Engineering faculty, industry experts, and researchers from the Center for Hazardous Materials Research. Finally, renowned guest lecturers from the Pacific Rim and Western Europe are brought to the MSEP classrooms by the University Center for International Studies, where they provide an international perspective on manufacturing economics, technology, and cultural difference. Thus, the MSEP course content varies from a rigorous master's level in the form of both cross-listed and specialized MSEP elective courses, to a high-level seminar that provides the student with an international perspective and a broad understanding of the significant issues currently facing the manufacturing industry.

In addition to its education and research components, the MSEP also develops a service component in order to provide a university "quick reaction capability" for its industrial partners. Strong faculty ties to industry are being developed through industry participation in university teaching and supervised student internships. The thesis level internships focus on solving specific company applied research problems. This is significantly increasing the pool of up-to-date knowledge and expertise that can be called upon in addressing important problems. The result is an effective model of industry-university collaboration, which is already fostering strong ties with both regional and national manufacturing industries.

A manufacturing systems engineering curriculum

The National Technological University (NTU), which offers courses and degrees via satellite, offers a Master of Science degree in manufacturing systems engineering. Totalling 33 semester credits, at least 5 credits of courses consisting of at least two additional courses must be taken in the areas of product and process design, manufacturing systems planning and control, quality control and reliability, quantitative methods, intelligent systems, software techniques and control theory. A minimum of 13 credits are required in the core courses covering design, modeling, control, management, and quality, with at least one course in each of these five areas.

A total of more than 125 in-depth graduate level courses are offered by NTU in the first list of topics described above, which can be taken for credit or audit. After completing 10 semester credits with grades of A's or B's, the student is eligible for an NTU Certificate of Completion. The MS degree usually takes about 1.5 years of full time graduate study or 3 to 5 years of part time study, depending on the course load.

Sample programs in manufacturing systems engineering

Area of Emphasis	Title	Credits
Advanced Automation	Core courses	
	Manufacturing systems design	3.0
	Modeling, simulation and optimization of manufacturing	3.0
	Vision, robotics, and planning	4.0
	Production planning and control	3.0
	Design for reliability	3.0
	Depth	
	Mechanical design for automated assembly	3.0
	Reliability engineering	3.0
	Breadth	
	Introduction to data structures	2.7
	Principles of engineering economics	3.0
	Elective	
	Computer graphics	2.7
	Material control applications	3.0
		33.4
Robotics and Control	Core Courses	
	Industrial robot design, selection, and implementation	3.0
	Digital simulation techniques	3.0
	Machine vision applications in mechanical engineering	3.0
	Production planning, scheduling, and inventory control	3.0
	Reliability	2.0
	Depth	
	Robotics: analysis and control	3.0
	Robot vision	3.0
	Breadth	
	Microprocessors	4.0
	Digital signal processing	3.0
	Elective	
	Probability and statistics for engineers	3.0
	Discrete optimization models and algorithms	3.0
		33.0

Words and Expressions

accredited *a.* 被认可的,质量合格的,经过认证的
board *n.* 板,委员会,研究会,(管理)局,厅
broad-based 包含广泛的
statistical process control 统计过程控制
rigorous ['rigərəs] *a.* 严格的,严密的
core 核心课程,必修课程
humanities and social sciences 人文与社会科学
communication skills 思想交流技能
teamwork 协作,协同工作
launch [lɔ:ntʃ] *v.* 创办,开办,提出,开始
recapture [ri:'kæptʃə] *n.* ;*v.* 取回,夺回,恢复,收复
participate [pɑ:'tisipeit] *v.* 参与,参加,分享,分担
auto industry 汽车制造业
initiative [i'niʃiətiv] *a.* 起始的,创造的;*n.* 第一步,着手,开创
recruit [ri'kru:t] *v.* 招收,吸收(新成员),充实,补充
managerial [mænə'dʒiəriəl] *a.* 管理的
embark [im'bɑ:k] *v.* 从事,着手,开始
multidisciplinary [mʌlti'disiplinəri] *a.* 包括各种学科的,多种不同学科的
faculty ['fæklti] *n.* (大学的)院、系,全体教师
supervision [sju:pə'viʒən] *n.* 监督,管理
senior executives 高级职员,高级主管人员
liaison [li'eizən] *n.* ;*v.* 联络,联系(人);协作 (with)
chaired professorship 讲座教授职位
complementary [kɔmpli'mentəri] *a.* 补足的,互补的,补充的
on-campus 在校内的
collaboration [kəlæbə'reiʃən] *n.* 合作,协作

conceptualization [kən'septjuəlai'zeiʃən] n. 形成概念,概念化
practicing ['præktisiŋ] a. 开业的,从业的,在工作中的
under the direction of 在…指导下
wrap up 结束
credit 学分
credit-hour 学分,学校或大学中计算学分的单位,通常指在一个学期内每周上课一小时
on-the-job 在职的
internship n. 实习生,实习
elective course 选修课程
course module 课程模块
renown [ri'naun] n. 名望,声望;v. 使有名望,使有声誉
renowned a. 有名望的,著名的
Pacific Rim and Western Europe 环太平洋的国家与地区和西欧
high-level 高阶层的,高级的
seminar ['seminɑ:] n. 讲座,讨论会
call upon 要求,需要
address n. 地址 v. 从事,忙于,处理
totalling n. 总计,共计,总和
eligible ['elidʒəbl] a. 符合被推选条件的,合格的

57. Manufacturing Research Centers at U.S. Universities

A variety of research centers devoted to manufacturing have been established at numerous universities across the country in an effort to maintain U. S. leadership in manufacturing technology. For instance, the University of Illinois, recognizing the importance of coordinated research activities, has initiated a program in manufacturing research and education to serve industry's needs. This program is being coordinated through the Manufacturing Research Center (MRC). The MRC is an industry-driven center of excellence in manufacturing research established to foster collaborative research initiatives between the university and industry. The Center conducts collaborative research projects with its member companies, educates students in the problems and issues of manufacturing, and provides a broader access for its members to the laboratories and programs of the university.

The research concentrates on four engineering sectors: electronics, automotive/vehicular, chemical, and machine tool. The Center funds two types of projects. Member companies designate that one-half of the funds it contributes be applied to research of a specific interest to that company. The results of the research from these company-designated projects are available on an exclusive basis to the company. The remaining funds from each company are employed collectively to support center-designated projects, the results of which are shared by all of the participating companies.

The MRC is also concentrating on robotics technology and manufacturing systems. The latter includes projects related to manufacturing systems design and control, integrated design and manufacturing, and management/systems integration. Robotics projects are in the areas of intelligent robotics, robotic vision, and path planning/control of robotic devices.

The Center for Manufacturing Engineering Systems (CMES) established at the New Jersey Institute of Technology (NJIT) has a similar goal as MRC—that of strengthening New Jersey's industrial base. Dedicated to the development and dissemination of advanced manufacturing knowledge, CMES is designed to serve industrial and govemment organizations. Activities focus on research and technology transfer. A particular emphasis of CMES is to aid small and medium-sized businesses in adapting advanced manufacturing technology to their needs. Funding has been received from a variety of sources, including the New Jersey Commission on Science and Technology, NJIT, and industry.

CMES-supported projects generally fall into one of three categories. The first project type, the most common, is the collaborative long-term project with industry, involving the application of computer-integrated manufacturing technology. For these projects, a company has a specific manufacturing systems problem it wants solved or a specific product it wants to develop. Most of these projects involve small to medium-sized firms and serve both a technology transfer and a research and development function.

The second type of project is the short-duration technology extension, collaborative project which aids industry in the selection and application of advanced manufacturing technology. For instance, a company seeking to purchase a CAD software package may come to CMES for advice about which device is most appropriate for this operation. The third project type involves typical university-based research on major manufacturing problems, with support from government and industry.

A substantial by-product of the CMES' activities is the provision of hands-on educational experience for the students who work on Center projects. Many of these students are enrolled in either the undergraduate and graduate degree programs in manufacturing engineering—the master's degree is the only such degree offered in New Jersey. Students earning this degree may select from several specializations, all of which share a common core of courses and call for completion of a thesis or project. Thesis and project topics, which usually involve applied research, are developed based on input from industrial advisors who keep NJIT faculty apprised of industrial priorities. NJIT has also teamed with community colleges to offer a new degree in manufacturing technology, a two-year CIM associates degree in applied science. Many students in this program complete part of their course work at NJIT, using the university's advanced laboratories, including the Manufacturing Automation Laboratory, a prototype CIM factory floor.

The Center for Manufacturing Productivity (CMP) is a collaboration between the UMass College of Engineering and the School of Management to provide a comprehensive program of assistance to small and medium-sized manufacturing companies in Western Massachusetts. CMP was founded in 1991 with a grant from the U. S. Small Business Association.

The CMP provides manufacturing businesses with comprehensive and integrated assistance in management and engineering through the expertise of UMass faculty in the School of Management and College of Engineering. Through on-site consultation and evaluation, the CMP staff, working with company management, will develop a detailed and cost-effective program specific to each company's needs. Among other strategies, CMP works with businesses to convert from defense to commercial manufacturing, develop computer-aided technologies and systems, modernize management structures, adapt to meet overseas product standards and adequately address export requirements of foreign markets, better utilize human resources, and improve production and control.

CMP offers a wealth of University expertise in engineering and management including CAD/CAM, strategic planning, design for manufacturing, market analysis, ISO 9000, polymer engineering, machining and grinding, total quality management, process control, human resource management, and software development.

Words and Expressions

concentrate ['kɔnsəntreit] *v.* 集中
vehicular [vi'hikjulə] *a.* 车辆的
designated project 指定项目
exclusive [iks'klu:siv] *a.* 排它的,排外的,专用的,独占的
intelligent robotics 智能机器人技术
short-duration 短期的
collaborative project 合作项目,合作课题
undergraduate and graduate degree 本科生和研究生学位
specialization [ˌspeʃəlai'zeiʃən] *n.* 特殊化,专门化,专业
call for 要求,提倡
associate degree 在美国大学中修完二年课程所获得的学位
comprehensive [kɔmpri'hensiv] *a.* 全面地,广泛地
integrated ['intigreitid] *a.* 综合的,完整的
UMass = University of Massachusetts
convert from defense to commercial manufacturing 将生产军工产品转为生产民用产品,军转民
adapt to 适合
strategic [strə'ti:dʒik] *a.* 战略的,重大的,关键的
planning 计划编制
process control 过程控制,工艺管理

58. A New Engineering Course

Imagine your job this next year was to teach a class of freshmen engineering students to swim—using a blackboard, a projector and a text.

Ridiculous, you say. By its very nature, swimming is a skill that must be learned by doing, preferably at least waist-deep in something wet, right?

Well, last year, University of Massachusetts Professors Corrado Poli, Keith Carver, Richard Giglio and staff took a good look at how their departments were being asked to teach engineers, and decided the same problem held true.

"What was wrong with the way we had been doing it was, at the end of a long, hard year, our freshmen knew no more about actual engineering than the day they walked through the door," Poli said. "Come June, they'd also lost a lot of the excitement they'd had about becoming engineers in the first place. Frankly, they were bored, and often, so were we."

According to Poli, the existing engineering curriculum had long rested on the theory that students needed a full year of theory, of learning skills in programming and spreadsheets and a host of other academic subjects, before they were prepared to undertake engineering.

Experience in the real world would tell us, however, that many a "better mouse trap" has been designed by someone with no engineering background at all—just someone who needed a mouse trap, Poli noted, "and at the same time, we were being told by people in industry that what they really needed, and weren't getting, was entry-level engineers who had good communication skills, the ability to work in a team, and who understood the need to design products for the ease of manufacturing."

The Department of Defense apparently agreed. Last year, UMass became a founding member of the Engineering Academy of Southern New England, a consortium of the region's major universities and colleges with engineering programs, who will share a $12M

Department of Defense "conversion" grant to recreate the engineering curriculum at all participating schools to meet the needs of industry. Major industries in the area will also participate in the public/private partnership by allowing students and faculty access to manufacturing sites and equipment. Poli, who was already collaborating on an advanced mechanical engineering book with John Dixon, a retired colleague, decided two of the chapters in that text were particularly suitable for addressing those concerns—and created a pilot program which was offered to a select group of second-semester freshman last spring.

"On the first day of class, I divided the class into teams of two and told them to find a design project, to choose any consumer product they wished, to analyze it, and make recommendations as to whether or not it should be redesigned, and if so, how. The only constraints were that the object had to have about 15 parts—and I advised that items with a lot of straight lines would be easier to draw when the time came to draw them. Only one of the 12 had any drawing experience whatsoever, and none had any computer-aided drawing experience," Poli said.

The students soon reported back with their choices, including a computer "mouse," an electric pencil sharpener, an electric fan, a computer disk, and a paint gun.

The challenge, and the fun, began.

How was their product made? Injection molding, die casting, stamping? Students had to learn the difference, and determine what was involved in each process.

How much manual assembly was required? And could it be reduced? To that end, students were required to produce actual assembly drawings, were introduced to Auto CAD, and simply learned to draw as they went along.

Drawings weren't the only form of communication required, no, progress reports, and presentations to the entire class were required as well. Abstracts were required, too, and after reading samples of good (and not so good) abstracts, students had to grapple with the age-old challenge of describing their own projects in no more than 200 words—and writing a good title, to boot.

The "results" writing assignment was also difficult, Poli said, as the class struggled to explain, in English, the same information they had been able to portray in graphs and tables with ease.

As they worked, team imbalances did show up, where one student put more energy into a given part of the project than another, Poli acknowledged, "but that happens in industry, too, and you have to learn to deal with it, to decide, are you going to do a bad job on a project because a colleague isn't pulling his weight?"

The final requirement: A five-minute oral presentation, made exactly as if it were being made to a board of directors in industry, detailing the new design recommendations—and with cost analysis required.

How'd they do?

"They were good, amazingly good," Poli said. "In fact, a colleague of mine who sat in on a couple of them had just finished sitting in on several upperclassmen's project presentations the week before, and he said these were better... by far."

Poli said even the students themselves were shocked by all that they'd accomplished: that by semester's end they were actually in a position to make concrete design suggestions, and, in fact, the actual manufacturer of the computer disk was now assessing that team's recommendations—which reduced projected assembly time by eight seconds per disk.

"It may not sound like much, but when you multiply that by

800,000 disks per week, you have some substantial savings to consider," Poli noted.

So successful was the pilot project that, this summer, additional members of the engineering faculty at UMass attended a workshop on the course, preparing them to teach it, and the student version will be offered to six sections of freshman this (Fall 1994) semester.

Second semester, graduates of that course will be offered mini-concentrations, taught in four-week sessions, in such subjects as Advanced Auto-CAD, computer programming, electrical engineering, materials science or process control. Faculty, meanwhile, will be busy continuing their own education in the new approach, which will require many more faculty to take part—"It is labor-intensive"—and also requires that they learn to teach cooperatively, and across the curriculum, themselves.

"Most faculty admit the idea is daunting," Poli said. "We're just not used to working together, not this way. There's no question that it was easier for us before, when we did it all our way, ourselves. But the people who have witnessed what happened last spring all agree: it will be worth it."

By September 1995, the College of Engineering hopes to have enough faculty "willing and able" to go on to offer the entire incoming freshman class—approximately 350 students in all—the "pilot" course, throwing them into the new curriculum stream en mass. That's also the year they hope to have the sophomore curriculum amended to build upon the work performed the year before.

By their junior year, it's now expected that future engineering students at UMass will be well on their way to... what? Performing like graduate students, working on, defending and fund-raising for projects, all on their own?

"Better than that," concludes Poli. "They'll be performing like engineers."

Words and Expressions

projector [prə'dʒektə] *n.* 投影仪
waist-deep 齐腰深的
in the first place 首先
existing [ig'zistiŋ] *a.* 现有的,目前的,现存的
academic subject 学科
undertake [ˌʌndə'teik] *v.* 从事,着手,进行
mouse trap 捕鼠器
entry-level 入门水平的
Department of Defense (美)国防部
consortium ['kən'sɔːtjəm] *n.* 合作,合伙,联合
$12M = 1 200 万美元
conversion [kən'vəːʃən] *n.* 转变,转换
address 致力于解决
pilot ['pailət] *a.* 引导的,导向的,(小规模)试验性的,试点的
whatsoever = whatever (语气比 whatever 强)无论什么,不管什么,诸如此类
computer mouse 计算机鼠标
injection molding 注射成型,注塑成型
die casting 压力铸造
go along 前进,进步,进行下去
progress report 进度报告,进展报告
presentation 讲述,展示,介绍
abstract [æb'strækt] *n.* 摘要
grapple ['græpl] *v.* 抓住,抓牢,握紧
grapple with a problem 抓问题,设法解决问题
age-old 古老的,久远的
to boot 并且,而且,加之,除此之外

portray [pɔ:trei] v. 描绘,描写,描述
with ease 熟练地,轻而易举的
imbalance [im'bæləns] n. 不平衡,不均衡
board of directors 董事会
sit in on 出席,(在会议或其他场合)旁听,旁观
upperclassman [ʌpə'klɑ:smən] n. 高年级学生
concrete a. 具体的,有形的,实际的
workshop ['wə:kʃɔp] n. 车间,专题研讨组,(专题)研讨会
mini-concentration 短期集中上课
labor-intensive 劳动密集型的
daunt [dɔ:nt] v. 威吓,使胆怯,使气馁
en mass 全部,全部地,整个地
amend [ə'mend] v. 修正,改进,改正
build upon 指望,依赖,建立于
defend [di'fend] v. 答辩,(为…)辩护
freshmen, sophomore, junior, senior (大学)一、二、三、四年级学生
graduate student 研究生

59. Online Web-Based Learning

Every student at a large university has experienced this: You take a seat among two or three hundred others in a large lecture hall; the lights go out; the slide projector comes on; and the lecture begins. Your thoughts may drift to your part-time job, your current relationship, the lab you have to make up. No matter how enthusiastic the lecturer is, no matter how passionate he or she is about the subject, you can safely retreat to your private zone because you know you won't be called upon. Soon the slides and the lecturer seem far away, and your eyes are getting very heavy...

"We call it 'darkness at noon,'" says Laetitia La Follette, associate professor at UMass Amherst. "We all work on our 'performances,' but no matter how hard we try, we know we're not reaching a lot of students. In fact, we know many of them don't even show up until there's a test." This phenomenon is not going away any time soon, says David Hart, executive director of the Center for Computer-Based Instructional Technology (CCBIT).

So, at a university with fewer teachers and larger classes, the question is: How do we make sure that the students in large survey courses are learning? One answer is the university's Online Web-Based Learning (OWL) system, which supplements the lecture hall and enhances the learning experience by providing interactive instruction that's designed to bring about illumination—at noon or any other time of day.

OWL, a combined learning management and "authoring" program that was created by CCBIT and faculty from the university's chemistry department, has many useful features. The authoring component enables instructors to write student assignments, including questions and feedback, and post them to the Internet. The instructors also add student rosters, so that enrolled students can get the assignments at any Internet-enabled computer—at school, home, or work. The learning management portion of the system keeps records, including quiz scores.

At its most basic, OWL allows extensive quizzing. Online students can get immediate feedback on each question they answer. If they get the answer wrong, they are told why, and can try the question again. Students do not need to do an entire assignment at once, but can work on a few questions and return later to do more. "On average," says Hart, "a student completes an online exercise three times before doing it correctly." This is in stark contrast to paper-based assignments, which allow students only one chance, and are graded by teaching assistants who may take days or weeks to give feedback.

In all, 28 departments and programs at UMass have placed homework assignments, tests and interactive instruction on OWL. Most of the 12,000 to 13,000 students who use OWL each year (some of these students are double-counted because they use OWL in multiple courses) are on campus, although OWL is used for a few distance eductaion courses.

OWL's primary support, CCBIT, is headed by Hart, who works with a group of approximately 10 faculty, staff and part-time students. This group's main function is to provide software development for continued enhancement of OWL, along with instructional design and development services to help professors effectively place their content in an interactive online environment.

CCBIT's support is very important in ensuring the quality and consistency of OWL instruction. Its job is to translate the faculty's ideas into something that can be effectively presented on a computer. The instructors bring subject knowledge and teaching experience. They know what the standards are, what students should get out of a course, and how best to explain the material. Then it's CCBIT's job to find effective ways to present that material online. This involves working with a team of graphic artists and programmers through several iterations of the material until it's done to everyone's satisfaction.

So how do we know OWL provides a better way to learn? That's where Alan Peterfreund of Amherst's Peterfreund Associates comes in. In the last four years, Peterfreund's firm has been an independent evaluator for many government sponsored education projects. In his work with UMass, he has been impressed with what he's seen: "What is interesting and profound is the degree to which these OWL projects influence what happens in the classroom. To the extent to which students come to class better prepared, teachers can do more exciting things. They can introduce interactivity, explore material at a greater depth, and introduce material at a higher level."

Peterfreund gauges the success of individual projects through two main assessments, the first of which involves collecting information—through surveys and interviews with students, faculty and teaching assistants—in order to get their perspectives on the effectiveness of the course. The second part is an attempt to collect information that demonstrates the success of the project—to make sure that learning objectives are met. Based on student scores and surveys and input from the instructors, CCBIT learns which methods are effective and which aren't. Thus many courses continually evolve, incorporating changes and new and better methods based on feedback.

The evaluation was positive, and students expressed great enthusiasm for the material. Even more gratifying to the project team was the demonstrated learning outcome. Students taking the course along with the new online modules scored an average of 20 percentage points better than did the control group, which used print materials or an older CD-ROM version of the modules.

Peterfreund notes that most OWL projects revolve around a central question:"How do you support innovative education at a large university where economics force you into large classes? Economics doesn't allow UMass to get away from large lecture classes, but that doesn't mean you need to be enslaved by old pedagogies."

The publisher Harcourt has licensed OWL to distribute some of the courses. This semester they will reach some 50 schools and over 25,000 students. All of this activity generates royalties for UMass, giving it a return on its investment, some $60,000 to date.

Words and Expressions

online web-based learning 基于网络的在线学习
lecture hall 大教室,讲演厅
go out 熄灭
slide projector 投影仪,幻灯片放映机
part-time job 兼职工作,非全日性工作
passionate ['pæʃənit] a. 充满热情的
retreat to 回到安静的地方,撤退到,退却
executive director 执行理事,常务董事
survey course 概论课
interactive instruction 交互式教学
illumination [iˌjuːmi'neiʃən] n. 照明,阐明,启发
authoring 撰写,编辑
instructor 任课教师
roster ['rəustə] n. 名册,人名册
quiz [kwiz] n. 小测验
in all 总共
double-counted 被重复计算
distance education 远程教育
along with 与…在一起
instructional [in'strʌkʃənəl] a. 教学的,教育的
profound [prə'faund] a. 意义深远的
to the extent 到…程度
gratify ['grætifai] v. 使满足
control group 对照组
CD-ROM 高容量只读存储器,光盘
revolve around 围绕,环绕
get away from 逃离,避免

enslave [in'sleiv] v. 奴役,束缚
pedagogy ['pedəgɔgi] n. 教学法,教育学
royalty ['rɔiəlti] n. 版权费,使用费

60. Laboratory Handbook for Technical Reporting (Ⅰ)

Introduction

The purpose of this handbook is to describe various aspects associated with acceptable laboratory practice. The standards established here will be the standards for all three of your laboratory courses in the mechanical engineering curriculum. The methods for graphical presentations or technical report writing are not the only acceptable methods. One has only to pick up two different textbooks to find two different styles of technical writing and two formats for making graphs. Certainly other methods are acceptable but unless a new method is specified you are to follow the techniques outlined in this handbook.

In the laboratory you will take measurements, analyze the results and draw conclusions based on your findings. Proper experimental methods, the opportunity to handle instrumentation and the ability to prepare competent technical reports form an important part of your engineering education. This handbook should help you develop those skills.

Before you go into the laboratory you should be properly prepared. You should know the experiment you are scheduled to perform. You should have read all the pertinent information that was assigned and you should have prepared your laboratory notebook. Adequate preparation is a must. It not only saves you time in the laboratory but it will make the laboratory a safer place for you, your lab partners and your colleagues.

And safety isn't the only factor. Proper preparation means there is less chance you will damage equipment. Some of the equipment you will use is expensive and some of it is the personal property of the faculty. Hopefully you will not damage equipment but if you do, report the damage immediately so it can be repaired. Generally the cost of equipment repairs is borne by the department, not by the student, unless the student is shown to be needlessly careless and this includes the failure to report the damage promptly. Remember that broken equipment is less than useless for everyone. It is frustrating, to say the very least, to take a piece of equipment off the shelf and find that it isn't working just when you need it.

Laboratory experiments are performed for a variety of reasons. The object may be to observe some phenomenon, measure some property or characteristic of some device or answer some question about the interrelationship of cause and effect between two or more parameters. The basic concept of experimentation or the experimental method has not changed significantly from the fundamental ideas expressed by Galileo back in the seventeenth century when his findings impelled him to challenge the teachings of the church regarding the earth as the center of the universe. The object of much of his work was the testing of hypotheses. The object or hypothesis was an expression about how things would react and he set about arranging an experiment to test this single idea. He formulated his conclusions based on the test results not on preconceived notions. His object was clear. His experiment was designed around this objective and his conclusions were directly related to it. The object of your laboratory work should be as clear to you as they were to Galileo.

Like Galileo your performance will be measured by evaluating your report. For Galileo his treatise on the subject of the motion of planets is found in DIALOGUE and has established him as the father of modern scientific experimentation. Certainly, there have been marvelous improvements in instrumentation since his time and the statistical tools and computer analysis techniques known today were unknown to him. However, his simple concept of doing an experiment properly and carefully and reporting the findings in a concise and coherent manner is still the foundation for experimental work today and is the premise upon which this handbook was developed.

Most of the material in this handbook is related to the presentation of technical information. This starts with the laboratory notebook and ends with the formal technical report. Read this information carefully for it pertains to all your laboratory work through your senior year.

Words and Expressions

outline ['autlain] v. 概括地论述
draw conclusion 得出结论
finding ['faindiŋ] n. 研究结果,试验结果,观察结果
competent ['kɔmpitənt] a. 有能力的,符合要求的,适当的
go into 进入,加入
lab 实验室
bear [bɛə] (bore,borne) v. 承担,负担,承受
needlessly ['ni:dlisli] ad. 不必要地
promptly ['prɔmptli] ad. 迅速地
frustrate [frʌs'treit] v. 挫败,阻挠,使感到灰心
cause and effect 因果
impel [im'pel] v. 推进,推动,驱使,促成,刺激
teaching n. 教学,讲授,(复数)学说,主义,教导,(宗教)教义

hypothesis [hai'pɔθisis] (*pl.* hypotheses) *n.* 假说,假设,前提
set about 着手,开始
preconceive [priːkən'siːv] *v.* 预想,事先想,事先做出的
notion ['nəuʃən] *n.* 观念,概念,看法,意见,见解
treatise ['triːtiz] *n.* (专题)论文
DIALOGUE 科学对话(伽利略著作的书名)
instrumentation [instrumen'teiʃən] *n.* 仪器,工具

61. Laboratory Handbook for Technical Reporting (Ⅱ)

Effective Report Writing

A key element in the advancement and refinement of engineering technology is the preparation of concise and informative reports which document the results of technical work. The value of any scientific research is highly dependent upon the quality and clarity with which its findings are presented since the report reflects the caliber of the work as well as the competence of the investigator.

Effective technical writing requires many of the skills and principles common to successful engineering: organization, analyzation, accuracy, efficiency, and consistency. Through common sense and experience, the art of clear and precise technical writing should become easy if you make a conscious effort to improve.

Three fundamental steps are essential to the development of a technical report:

(1) Analyze the objective. The objective governs the direction in which the report proceeds and the scope of the objective limits the report content. The objective is the goal towards which the writer's efforts are directed. Keep the objective in mind at all times.

(2) Assemble the pertinent information. Information is necessary to support development in the report. Sources of information may include: documentation, experimentation, observation, and logical thinking. Only material which is relevant to the report objective should be included.

(3) Prepare a working outline. The outline delineates the magnitude and direction of the writer's task. Design the structure of a report to ensure a logical and fluid progression towards a definitive conclusion.

Once you've organized an outline, the actual writing process may begin. The writing should generally proceed as follows:

(1) Prepare a rough draft putting information down in your own words following the format of your outline.

(2) Refine the text by rearranging material for better clarity, removing ambiguities and correcting for misspellings and errors in grammar. This requires the writer to read and reread his words while critically examining each sentence. Read each sentence carefully to understand its meaning. The sentence should be clear and concise and subject to a single interpretation. Take care to ensure each sentence says exactly what you mean to say, each paragraph works with a single idea, and that the report contains only material that is pertinent to the objectives.

(3) Prepare a final draft so that the text material is legible. Type the report if possible to ensure a neat, clean presentation. Prepare the tables, graphs and figures in their final form along with all other supplemental information. Collate the material properly.

This procedure holds for a variety of technical reports. But keep in mind when you are writing a report, the type of report you wish to write and the audience for whom it is being written. The following section describes some of the different types of technical reports.

Any efficient communication must be designed for the needs and understanding of a specific audience. The report must be fine-tuned to the level of the reader and fit the requirements of the report objective. Several types of technical reports exist and the structure as well as the style of each report will vary accordingly.

Internal Publications

Internal implies that the report is being produced for the benefit of the company, institution, or any organization which has control over the work being reported. Internal reports are often restricted to circulation within the system and can range from relatively informal memoranda to highly technical and detailed reports.

Memorandum: A short report will often be written as a memorandum which consists of a brief and direct transferal of information, free of background material and lengthy development. The style is often conversational and directed to a co-worker or immediate supervisor. Emphasis is on the objective and the conclusion.

Detailed Report: A detailed report follows a formal development and serves as a solid presentation of some type of technical work. The format complies with the standards and methods of the institution for which it is written. Extensive detail and proprietary data are often documented in an appendix; these reports serve to record such information for future reference.

External Publications

External publications are typically an overview containing important findings and conclusions for the benefit of a broad technical community. The requirements for a paper submitted to the American Society of Mechanical Engineers (ASME) are presented in Appendix A. Note that many of the report elements presented in the following text adhere to these requirements.

APPENDIX A
Requirements for Papers Submitted to the ASME

Manuscript Format

· Manuscripts should be submitted in final form to the Editor. Each manuscript must be accompanied by a statement that it has not been published elsewhere nor has it been submitted for publication elsewhere. A paper which would occupy more than 6 pages of the Journal will be returned to the author for abridgment.

· The author should state his business connection, the title of his position, and his mailing address. A short abstract (50 to 100 words) should be included on the first page immediately preceding the introductory paragraph of the paper.

· Five copies of the manuscript are required.

Use of SI Units

· All manuscripts submitted to ASME must use SI (Metric) Units in text, figures, or tables. In addition to SI units, English units may be included parenthetically.

Mathematical Expressions

· All mathematical expressions should be typewritten. Care should be taken to distinguish between capital and lower-case letter, between zero (0) and the letter (o), between the numeral (1) and the letter (l), etc.

· Numbers that identify mathematical expressions should be enclosed in parentheses. Numbers that identify references at the end of the paper should be enclosed in brackets. Care should be taken to arrange all tables and mathematical expressions in such a way that they will fit into a single column when set in type. Equations that might extend beyond the width of one column should be rephrased to go on two or more lines within column width.

Length

· The text of an ASME paper normally should not exceed 6000 words (6 printed pages in a journal) or equivalent. A Technical Brief or Brief Note should not exceed 1500 words or equivalent. In computing equivalence, a typical one-column figure or table is equal to 250 words. A one-line equation is equal to 30 words. The use of a built-up fraction or an integral sign or summation sign in a sentence will require additional space equal to 10 words.

· Title of papers should be brief.

Words and Expressions

refinement [ri'fainmənt] n. 精炼,提纯,细致的改进,改进的设计
caliber ['kælibə] n. 质量
competence ['kɔmpitəns] n. 能力,胜任
common sense 常识
conscious ['kɔnʃəs] a. 有意识的,故意的
progression [prə'greʃən] n. 进步,进展,前进
definitive [di'finitiv] a. 最后的,确定的
ambiguity [æmbi'gjuiti] n. 意义不明确,模棱两可的话
misspelling [mis'speliŋ] n. 拼写错误
interpretation [intə:pri:'teiʃən] n. 解释,说明,诠释
legible ['ledʒəbl] a. 清楚的,明了的,易读的
collate [kɔ'leit] v. 整理,排序,分类
hold for 适用
publications 出版物,发行物

memorandum [meməˈrændəm] n. 备忘录,便笺,便函
solid a. 可靠的,坚固的
comply with 遵守,服从
appendix [əˈpendiks] n. 附录
abridgment [əˈbridʒmənt] n. 缩短,节略
business connection 工作单位
introductory [intrəˈdʌktəri] a. 引言的,导言的
manuscript [ˈmænjuskript] n. 稿件
parenthetically [pærənˈθetikəli] ad. 顺便地,作为插句
capital and lower-case letter 大写和小写字母
parenthesis [pəˈrenθisis] n. (pl. parentheses) 插句,括弧,圆括号
rephrase [riːfriz] v. 改用别的措词表达
equivalent [iˈkwivələnt] a. 相当的,相等的; n. 同等物,相等物
built-up fraction 并排的分数式
integral sign 积分号
summation sign 求和符号

62. Laboratory Handbook for Technical Reporting (Ⅲ)

Technical Report Elements

Abstract

The Abstract is a condensed statement of the important information contained in the complete report. It is the epitome of a summary. It stresses the objective and conclusions. An Abstract allows the reader to survey the purpose, content, and conclusions of a report quickly. The two most important requirements of an abstract are that it be concise and informative. To accomplish this, the abstract is usually written last.

Introduction

The primary function of an Introduction is to let the reader know the importance of the work and to clearly define the objective. Once this is stated, a brief plan of development should follow. A well-constructed Introduction should stimulate reader interest and summarize the contents of the report. Background information of a theoretical or historical nature may be warranted to support this preliminary information. As you would expect, the Introduction is the section that introduces the work to the reader. The beginning of the Introduction usually explains the problem and the objective of the report. Journalists are taught to answer the five W's: who, what, where, when and why. It is probably a good idea to keep these in mind when writing your Introduction and to answer those W's that are pertinent to your objective. For example, the WHO might be the names of previous investigators that you found in a literature search. The WHAT would be the problem statement. The WHY might be pertinent applications of your results. The WHERE and WHEN are obvious.

The Introduction is used to acquaint the reader with the material of the report. As part of this presentation it is advisable to state some of the important principles of the work and enumerate assumptions.

Analysis

The Analysis section is used to develop a pertinent theory based on the basic principles that explain the phenomenon you are investigating. Most experimental studies involve the interaction of a variety of complex influences and subtleties. The purpose of the analysis is to remove the mask of complexity and expose the underlying facts. It is a process of systematic thinking, combining logical assumptions with basic principles to develop a relationship that explains your results. This relationship is usually the hypothesis that is the subject of the report. The experiment is the study of this hypothesis to test if your assumptions and logic are correct.

The Analysis is usually interspersed with equations. It is not simply a series of equations devoid of explanatory material. The explanation of technical material is naturally associated with mathematics. Assumptions, which are expressed in words, are transformed into their mathematical equivalents. Basic principles are also expressed in mathematical terms and are combined with the assumptions to develop the hypothesis. Intermediate steps showing the algebra and calculus, while necessary to the development of the hypothesis, are not shown. However, your presentation should be complete enough that a peer could duplicate your work. It is frustrating to see the expression, "it can readily be seen that" between two equations which bear no apparent relationship one to the other. If some real detail is necessary to fully explain a particular point but is extraneous to your basic presentation, then this work belongs in an appendix.

Equations must be presented clearly with explanatory material relating the equation to the remainder of the report. Symbols should be defined when they are first introduced. All the symbols in an equation must be defined. However, it is not necessary to redefine terms once they have been presented. If the report contains a number of unfamiliar symbols, give a nomenclature section.

Procedure

This section describes the apparatus and details the experimental procedure for taking measurements. In this section, you must explain what was measured and how you measure it. You should provide sufficient detail so that the experiment can be replicated using the same or equivalent equipment. Drawings showing the setup are often useful. They can be an aid in describing certain measurements and they should show the interconnections of the various instruments.

The Procedure does not contain results. You can explain that 20 separate tests were performed. You can say that the means and standard deviation were calculated, but you do not give the numerical values. These values are presented in the Results Section.

Discussion of Results

Results are the facts. They are the data you collected and the data you calculated. Means, standard deviations, confidence intervals and errors are all results.

Present the results in a logical and concise fashion. You can place sample calculations in this section. But if you want detail and an explanation of a series of extraneous calculations, then use an appendix. In general, the detailed calculations of the experimental errors are best placed in an appendix unless the analysis of the errors is the object of the report.

Do not transcribe your raw data. These are the numbers you recorded from your experimentation. A Xerox or carbon copy of these data should be in the appendix forming the last page of your report. Thus, it is important to keep a neat, clear and informative laboratory notebook, and all your lab partners could have the same identical last page of their report.

In the Procedure, you explained how and what were measured. Now you give the results. The results are the facts; given the same raw data, the reader should get the same results. Repeating the experiment should give similar results. But even when the results are identical, readers may interpret the results differently weighing certain information more heavily. These interpretations of the results are called conclusions.

Conclusions

It is interesting that, given the same results, two people can draw two different conclusions, and neither conclusion is necessarily incorrect. That is not to say that any conclusion is correct but that a conclusion is personal; it is your interpretation of the results and is subjective. However, the conclusion should relate to the objective of the report.

Students hesitate to make conclusions for fear of being wrong. "This method of determining the coefficient of friction was a reasonably good way of obtaining fairly accurate results," says nothing. It straddles the issue and avoids being wrong. It is better to be decisive when the results warrant a decision.

Some legitimate conclusions are:

(a) This experiment showed that the coefficient of static friction between aluminum and brass is not a simple value but can vary by as much as 50%.

(b) This experiment showed that it is not necessary to use sophisticated or expensive equipment to obtain accurate results.

(c) For an experiment stressing precision, the equipment was unusually crude. No wonder the results had such variability. Better equipment would have given more precise answers.

All these conclusions may be valid; it depends upon the results. Remember that conclusions are not facts. They are your interpretation of the facts, and these interpretations should pertain to the objective of the report. They should bring your report to a sensible finish.

Words and Expressions

epitome [i'pitəmi] *n.* 梗概,概括,缩影,集中表现
introduction 引言,导论,前言
stimulate ['stimjuleit] *v.* 激发,刺激
preliminary [pri'liminəri] *a.* 初步的,预备的,开端的

investigator 研究的,研究人员
literature search 文献检索
enumerate [i'nju:məreit] v. 数,列举,计算
experimental study 实验研究,实验
subtlety ['sʌtliti] n. 微妙,微细,敏锐
underlying [ʌndə'laiŋ] a. 下面的,潜在的,根本的
subject 题目,主题
assumption [ə'sʌmpʃən] n. 假定,设想
intersperse [intə'spə:s] v. 散布,散置
intersperse with 用…点缀着,不时用…打断
devoid [di'vɔid] a. 缺乏,没有(of)
peer 同等的人,同行
extraneous [eks'treinjəs] a. 外部的,附加的,无关的,不重要的
remainder [ri'meində] n. 剩余部分
nomenclature [nəu'menklətʃə] a. 名词,术语,术语表
apparatus [æpə'rætəs] n. 器械,设备,仪器
mean and standard deviation 平均值和标准差
confidence interval and error 置信区间与误差
transcribe [træns'kraib] v. 抄写,记录
Xerox or carbon copy 复印或复写本,副本
weighing [weiiŋ] n. 权衡,权重
hesitate ['heziteit] v. 犹豫,不愿
coefficient of friction 摩擦系数
straddle ['strædl] v.;n. 对…不表态,骑墙,观望
legitimate [li'dʒitimit] a. 合法的,正规的,合理的,真实的
sensible ['sensəbl] a. 切合实际的,合理的,有判断力的

Glossary

A

3-phase ac 三相交流电

abbreviation [əˌbriːviˈeiʃən] n. 缩写,缩写词

ablative [ˈæblətiv] a. 烧蚀的,脱落的;n. 烧蚀材料

abrasion [əˈbreiʒən] n. 擦伤,磨损,磨耗

abrasive [əˈbreisiv] n. 磨料,研磨剂;a. 磨料的,磨蚀的

abridgment [əˈbridʒmənt] n. 缩短,节略

absolute liability 绝对责任,绝对赔偿责任

absorber [əbˈsɔːbə] n. 减震器,缓冲器,阻尼器

abstract [æbˈstrækt] n. 摘要

accelerated testing 加速试验

access cover 进出口盖,舱口盖,检修盖

accessibility [ækˌsesibiliti] n. 可接近性,易维护性,检查,操作

accessory [ækˈsesəri] a. 附属的,附带的,n. 附属品,附属装置

accountant [əˈkauntənt] n. 会计,出纳

accreditation [əkrediˈteiʃən] n. 任命,鉴定,认证

Accreditation Board for Engineering and Technology (ABET) 工程技术认证委员会

accredited a. 被认可的,质量合格的,经过认证的

accumulated error 累积误差

accumulation [əkjuːmjuˈleiʃən]n. 积累,累加,累积

acoustics [əˈkuːstiks] n. 声学

acquired [əˈkwaiəd] a. 已得到的,已获得的

acronym [ˈækrənim] n. (首字母)缩写词,简称

activate ['æktiveit] v. ;n. 开动,启动,驱动,激发
actuator ['æktjueitə] n. 执行机构,致动器
adaptable [ə'dæptəbl] a. 可以适应的,能改变的,通用的
adaptation [ə'dæpʃən] n. 适应,配合,匹配
adaptive [ə'dæptiv] a. 适合的,适应的,自适应的
adaptive control 自适应控制
additive ['æditiv] n. 添加剂,外加物
address v. 专注于,致力于,从事于
adherence [əd'hiərəns] n. 粘着,粘附,粘着力
adsorb [æd'sɔ:b] v. 吸附,吸取
adversely ['ædvə:sli] ad. 有害地,不利地,相反地
aeroengine ['ɛərəu'endʒin] n. 航空发动机
aforementioned [ə'fɔ:'menʃənd] a. 上述的,前面提到
Age of Stone 石器时代
algorithm ['ælgəriðəm] n. 算法
aligning [ə'laignin] n. 校直,直线对准
alleged [ə'ledʒd] a. 被断定的,被当作确实的,声称的
allowance [ə'lauəns] n. 公差,允许量,加工余量
allude [ə'lju:d] v. (间接)提到,引证,指...说
alphanumeric ['ælfənju'merik] a. 字母数字(混合编制)的,n. 字母数字符号
alternating current 交流电
alternative [ɔ:'l'tə:nətiv] n. 抉择,可供选择的方法(事故);a. 选择性的
alumina grinding media 氧化铝磨料
amend [ə'mend] v. 修正,改进,改正
Amonton´s law 阿蒙登摩擦定律
analogous [ə'næləgəs] a. 类似的,相似的,可比拟的
ancillary [æn'siləri] a. 辅助的,附属的;n. 辅助设备
anneal [ə'ni:l] v. ;n. (使)退火,(加热后)缓冷
annular ['ænjulə] a. 环形的,环的

antifriction bearing 减摩轴承,滚动轴承
anti-backlash 消隙,消除间隙
apart [ə'pɑ:t] ad. 相隔,相距,分开,拆开
aperture ['æpətjuə] n. 孔,壁孔,孔隙,孔径
apparatus [æpə'rætəs] n. 器械,设备,仪器
appendix [ə'pendiks] n. 附录
application 应用程序,应用软件
application server 应用服务器
appointment [ə'pɔintmənt] n. 指定,家具,车身内部装饰
armature ['ɑ:mətjuə] n. 电枢,转子
articulate [ɑ:'tikjuleit] v. 铰接,活动连接
articulated [ɑ:'tikjuleitid] a. 铰链的,有活动关节的,关节式的
articulated robot 关节型机器人
artificial intelligence 人工智能
artificial neural network 人工神经网络
as shown 如图所示
ASCE 美国土木工程师协会
asperity [æs'periti] n. 粗糙,凹凸不平
assembled a. 装配的,组合的,已装配好的
assembly [ə'sembli] n. 装配
assessment [ə'sesmənt] n. 评估,评价,鉴定
assumption [ə'sʌmpʃən] n. 假定,设想
asynchronous [ei'siŋkrənəs] a. 不同时的,异步的
attachment [ə'tætʃmənt:] n. 附件,附属装置
attribute ['ætribju:t] n. 属性,特征,标志,象征
attribute M to N 认为M是由于N引起的
audit ['ɔ:dit] n. (大学生)旁听(课程),审计,查账
authoring 撰写,编辑
auto industry 汽车制造业
automaker 汽车制造商

automatic [ɔ:tə'mætik] n. 自动机械,自动装置; a. 自动化的
automatic screw machine 自动螺丝车床
automatic stop 自动停机,自动停车装置
automatically guided vehicle 自动导引小车,自动制导车辆
automaticity [ɔ:'tɔmətisəti] n. 自动化程度
auxiliary [ɔ:g'ziljəri] a. 辅助的
axial ['æksiəl] a. 轴的,轴向的
axiom ['æksiəm] n. 公理,规律,原则
axiomatic [ˌæksiə'mætik] a. 公理的,公理化的
axle ['æksəl] n. 轮轴,车轴

B

backlash ['bæklæʃ] n. 间隙,齿间隙
back-pressure 背压,反压
ball nut 滚珠螺母,滚珠螺母组件(由滚珠螺母、钢球、循环机构和附件所构成的组件)
ballscrew ['bɔ:lskru:] n. 滚珠丝杠
bandwidth ['bændwidθ] n. 带宽,频带宽度
bar stock 棒料
barring ['bɑ:riŋ] prep. 除...以外,不包括...
base material 基体材料,母料
base plate 基础板,底板
basic major diameter 外径的基本尺寸,公称直径
batch production 成批生产,批量生产
batch program 批处理程序
bathtub ['bɑ:θtʌb] n. 澡盆,浴缸
be equipped with M 装备有M,安装有M
bearing ['bɛəriŋ] n. 轴承,支座,支架
bearing surface 支承面
bench model 台式

bench space 工作台面积

bend [bend] n.;v. 弯曲,弯曲部

bespoke [bi'spəuk] a. 专做订货的;n. 预订的货

bill of material 材料表,材料清单,物料清单

bind [baind] (bound,bound)v. 捆,结合,凝固,粘合

bipolar transistor 双极性晶体管,场效应晶体管

blade (无心磨床上的)托板

blueprint ['bluːprint] n. 蓝图,设计图

bolt [bəult] n. 螺栓;v. 用螺栓固定

bonding ['bɔndiŋ] n. 连(焊,胶,粘)接,粘结剂

bookkeeping ['bukiːpiŋ] n. 簿记,记账

boot up 引导,使系统开始工作

boring ['bɔːriŋ]n. 镗孔,镗削加工

bottom line 最后的结果,底线

boundary lubrication 边界润滑

bracket ['brækit] n. 括号

brainpower 智能,智慧,科技人员

brake [breik] n. 制动器,制动装置;v. 制动,刹车

breach [briːtʃ] n. 破坏,违反,不履行

breakage ['breikidʒ] n. 破损,断裂,损坏

breed [briːd] v. 生育,复制;n. 品种,种类

brittle ['britl] a. 易碎的,脆的

browse [brauz] v. 浏览,翻阅

building block 构件,组成部件,标准部件

bump [bʌmp] v.;n. 撞,碰撞,冲击

burn out 烧坏,烧掉

burr [bəː] n. 毛刺

by-product 副产品,出乎意料的结果(或效果)

C

calculus ['kælkjuləs] *n.* 计算,微积分
call on 访问(内存储单元)
calling ['kɔːliŋ] *n.* 职业,行业,名称
cam [kæm] *n.* 凸轮
camshaft ['kæmʃɑːft] *n.* (机械)凸轮轴
capacitance [kə'pæsitəns] *n.* 电容,电容量
capacitive [kə'pæsitiv] *a.* 电容的
capital investment 资本投资,基本建设投资
capital letter 大写字母
capitalize on 利用
carbide ['kɑːbaid] *n.* 碳化物,硬质合金
carriage ['kæridʒ] *n.* 溜板(在床身上使刀具作纵向移动的部件。一般由刀架、床鞍、溜板箱等组成)
carrier ['kæriə] *n.* 载体,承载部件
Cartesian coordinate system 笛卡尔坐标系,直角坐标系
case harden 表面硬化,表面淬火
casting ['kɑːstiŋ] *n.* 浇铸,铸件
catalog ['kætəlɔg] *n.* 目录,一览表,(产品)样本
categorize ['kætigəraiz] *v.* 分类,分门别类
CATIA (computer aided three dimensional interactive application) 计算机辅助三维交互式应用
cause and effect 因果
cavity ['kæviti] *n.* 空腔,(铸造)型腔,室穴,腔体,凹处
CD-ROM 光盘只读存储器
centerless ['sentəlis] *a.* 无心的,没有心轴的
centerless grinder 无心磨床
centerpiece ['sentəpiːs] *n.* 主要特征
centralized 集中的

ceramic insert 陶瓷刀片

ceramics [si'ræmiks] n. 陶瓷

certified ['sə:tifaid] a. 持有证明书的,有书面证明的,经过检定的

chaired professorship 讲座教授职位

changeover ['tʃeindʒəuvə] n. 转换,(生产方法,装备等的)完全改变,调整

chatter ['tʃætə] n. 振动,颤震现象

check against 检查,核对

chip [tʃip] n. 切屑,金属屑,芯片

chuck [tʃʌk] n. 卡盘,(电磁)吸盘

chucking lathe 卡盘车床(床身较短,无尾座的卧式车床)

CIM 计算机集成制造

circuitry ['sə:kitri] n. 电路,线路,电路图,电路系统

clamping ['klæmpiŋ] n. 夹紧,固定

class of fit 配合级别,配合类别

cleanliness ['klenlinis] n. 清洁度,洁净

clerical ['klerikəl] a. 书写的,文书的,事物性的

click on 在⋯上单击

clutch [klʌtʃ] n. 离合器;v. 使离合器接合

CNC = computer numerical control 计算机数字控制

coaxial [kəu'æksiəl] a. 同轴的,共轴的

coaxial cable 同轴电缆

cobalt ['kəubɔ:lt] n. 钴

coded instruction 编码指令

coefficient of friction 摩擦系数

coherent [kəu'kiərənt] a. 连贯的,有条理的,紧凑的,连贯的,清晰的

coherent system 单调关联系统(指系统具有单调可靠性结构函数,且系统中任一部分的状态都与系统状态有关),相干系统

coincide [kəuin'said] v. 重合,一致

cold forming 冷成型,冷态成型,冷模压

cold-headed 冷镦的
collaboration [kəlæbə'reiʃən] n. 合作,协作
collaborative [kə'læbəreitiv] a. 合作的,协作的
collaborative project 合作项目,合作课题
collet ['kɔlit] n. 夹头,有缝夹套
collision [kə'liʒən] n. 碰撞,冲突
color display 彩色显示器
commentary ['kɔməntəri] n. 解说词,注释
communication channel 信息通道
communication skills 思想交流技能
compass ['kʌmpəs] n. 圆规
compatibility [kəmpætə'biliti] n. 相容性,兼容性,适合性,一致性
compatible [kəm'pætəbl] a. 相容的,兼容的,可共存的,相适应的
competent ['kɔmpitənt] a. 有能力的,符合要求的,适当的
complement ['kɔmplimənt] v. 补充,互补,补充物
complementary [kɔmpli'mentəri] a. 补足的,互补的,补充的
complete interchangeability 完全互换
compliance [kəm'plains] n. 依从,顺从,柔度
component part 组合零件,部件,组成部分
composite ['kɔmpəzit] a. 合成的,复合的;n. 复合材料,混合料
comprehensive [kɔmpri'hensiv] a. 全面地,广泛地
compress [kəm'pres] v. 压缩,缩短
compromise ['kɔmprəmaiz] n. ;v. 折衷,平衡,综合考虑
computer simulation 计算机模拟,计算机仿真
computer-aided process planning (CAPP) 计算机辅助工艺过程设计
computer graphics 计算机制图,计算机图形学
computer-guided 计算机导向的
computer-integrated manufacturing 计算机集成制造
conceive [kən'si:v] v. 设想,想象,表现
concentrate ['kɔnsəntreit] v. 集中

concentric [kən'sentrik] a. 同中心的,同轴的
concentricity [ˌkɔnsen'trisiti] n. 同心,同心度
conceptual [kən'septjuəl] a. 概念上的
conceptual stage 总体设计阶段,方案设计阶段
conceptualization [kən'septjuəlai'zeiʃən] n. 形成概念,概念化
concomitant [kən'kɔmitənt] a. 伴随的,随...而产生的
concrete a. 具体的,有形的,实际的
concurrent [kən'kʌrənt] a. 同时发生的,并行的;n. 同时发生的事件
condense [kən'dens] v. 浓缩,压缩,简要叙述
conductivity [kəndʌk'tiviti] n. 传导率,传导系数
confidence interval 置信区间
configurable 结构的,可配置的
configuration [kənfigju'reiʃən] n. 外形,构造,配置,结构
configuration design 配置设计,结构设计
configure [kən'figə] v. 配置,设定,使...成形
confine [kən'fain] v. 限制在...范围内;n. 区域,范围
conformance [kɔn'fɔːməns] n. 相似,一致性,适应性
consistency [kən'sistənsi] n. 一致性,连续性
consolidate [kən'sɔlideit] v. 巩固,统一,整理
consortium ['kən'sɔːtjəm] n. 合作,合伙,联合
constituent [kɔn'stitjuənt] a. 组成的 n. 组成部分,构成
constrained [kən'streinid] a. 限定的,受约束的
consumable [kən'sjuːməbl] a. 可消耗的;n. 消耗品
consumer product 消费品
contaminant [kən'tæminənt] n. 污染物,杂质
contest [kən'test] v. ['kɔntest] n. 争论,辩论,比赛
continuous path machining 连续轨迹加工
contour ['kɔntuə] n. 轮廓,形状,外形
contour following 轮廓跟踪(技术,方法)
contouring system 轮廓系统

contract ['kɔntrækt] n. 合同,契约
contract out 订合同把工作包出去
contractual [kən'træktjuəl] a. 契约的,合同的
control group 对照组
convergence [kən'və:dʒəns] n. 收敛,集中,减小
conversion [kən'və:ʃən] n. 转变,转换
converter [kən'və:tə] n. 转换器,变换器,变流器
conveyance [kən'veiəns] n. 运送,传送
coolant ['ku:lənt] n. 冷却液,切削液,乳化液
coordinate grid 坐标网
coordinate measuring machine 坐标测量机
core course 核心课程,必修课程
cornerstone ['kɔ:nəstəun] n. 基础,基石
corollary [kə'rɔləri] n. 推论,必然的结果
correlation [kɔrə'leiʃən] n. 相互关系,相关性
corrosion [kə'rəuʒən] n. 腐蚀,侵蚀,锈蚀
cost effective 有成本效益的,节省成本的
cost estimating 费用概算,成本估计
cottage industry 家庭手工业
counsel ['kaunsl] v. 劝告;n. 讨论,辩护律师,法律顾问
counseling ['kaunsəliŋ] n. (对个人,社会以及心理等问题的)咨询服务
counteract [kauntə'rækt] v. 抵消,中和,阻碍
counterclockwise ['kauntə'klɔkwaiz] a.;ad. 逆时针方向的
counterpart ['kauntəpɑ:t] n. 相似物,对应物,另一个具有相同功能或特
　　　　　　　　　　　　　　色的人或物
countereaction [kauntə'rækʃən] n. 反作用(力),抵消,中和,对抗作用
couple ['kʌpl] v. 使联在一起,联接,力偶
coupling ['kʌpliŋ] n. 耦合,联轴器,连接器
course module 课程模块
co-existent 同时共存的

cram [kræm] v. 填满,勉强塞入
crank [kræŋk] n. 曲柄
credibility [kredi'biliti] n. 可信性
creep [kri:p] n. 蠕变,徐变
creep feed 缓进给
creep feed grinding 缓进给磨削
crib [krib] n. 木笼
crisp [krisp] a. 清新的,明快的
critical factor 关键因素
cross section 横截面
cross slide 横刀架
cross-functional 具有多种职能的,由多种不同职务人员组成的
cross-training 交叉培训,交替培训
crucial ['kru:ʃəl] a. 决定性的,关键的,紧要关系的
crush [krʌʃ] v. ;n. 压碎,碾碎,碎碎
crystalline ['kristəlain] a. 结晶(质)的,晶状(态,体)的
culminate ['kʌlmineit] v. 达到顶点,结束
culminate in 以...而终结,以...而达到顶峰
culmination [kʌlmi'neiʃən] n. 顶点,最高潮,极点
current 电流
current loop 电流回路
current version 现在的型式,目前的样式
curriculum [kə'rikjuləm] (pl. curricula) n. 课程,学习计划
cushion ['kuʃən] n. 缓冲器;v. 缓冲,减振
cylinder ['silində] n. 圆柱,柱体,气缸,油缸
cylindrical [si'lindrikəl] a. 圆柱的,圆柱形的
cylindrical grinding 圆柱面磨削,外圆磨削

D

damp [dæmp] n. ;v. 阻尼,衰减

dampen ['dæmpən] v. 抑制,使衰减,阻尼,减震,缓冲

data processing 数据处理

database ['deitbeis] n. 数据库

day in and day out 天天,一天接一天

debug [di:bʌg] n. (程序)调整,排除故障

deburr [di'bə:] v. 去毛刺

deceleration [di:selə'reiʃən] n. 减速(度),降速

decompose [di:kəm'pəuz] v. 分解

dedicated ['dedikeitid] a. 专用的

dedicated automation 专用自动化设备

deenergize v. 切断,断开,释放(继电器,电磁铁等),去(解除)激励

default [di'fɔ:t] n. 系统设定(预置)值,默认(值),缺省(值);v. 默认

defectively [di'fektivli] ad. 有缺陷地,缺乏地

deficiency [di'fiʃənsi] n. 缺乏,不足,缺点,不足之处

definitive [di'finitiv] a. 最后的,确定的

deflect [di'flekt] v. 偏移,弯曲,下垂

deflection [di'flekʃən] n. 偏移,偏转,弯曲,挠度

deformation [di:fɔ:'meiʃən] n. 变形

delegate ['deligit] n. 代表;v. 委派,授权

delicate ['delikit] a. 精密的,精巧的,灵敏的

delineate [di'linieit] v. 描出...外形,画出...轮廓,描写,叙述

delivery process 交付过程

Delphi method 德尔菲法(这种方法是通过通信填表的方式,广泛征求技术人员和专家对某项新技术发展的意见,将意见整理分类后,对不同意见和分歧意见,再进行第二轮征询,如此重复数次,最后将基本一致的意见整理出来公布于众)

dependable [di'pendəbl] a. 可靠的,可信任的

depict [di'pikt] v. 描述,描写

depression [di'preʃən] n. 沉降,凹陷

derivative [di'rivətiv] n. 衍生物;a. 派生的,衍生的

derive [di'raiv] v. 从...得到,衍生出,获得,引伸出

descending [di'sendiŋ] n. ;a. 下降(的),下行(的),递降(的)

design change 设计变更

descriptive geometry 画法几何

design flaw 设计缺陷,设计瑕疵

design for manufacture (DFM) 面向制造的设计

designated project 指定项目

designation [dezig'neiʃən] n. 牌号,命名,规定,目标

detail v. 清晰地说明;详细叙述

detail drawing 零件图,详图

deterioration [ditiəriə'reiʃən] n. 变质,退化,恶化,变坏

deterministic [ditə:mi'nistik] a. 确定性的

detrimental [ˌdetri'mentəl] a. 有害的,不利的

devastating ['devəsteitiŋ] a. 破坏性的

developed country 发达国家

deviation [di:vi'eiʃən] n. 偏差,偏移,差异,误差

devoid [di'vɔid] a. 缺乏,没有(of)

diagram ['daiəgræm] n. 图表,示意图,特性曲线

dial ['daiəl] n. 刻度盘,仪表面,指针

dial indicator 刻度指示盘,带刻度盘的指示器

diameter-pitch combination 直径-螺距组合

dictate ['dikteit] v. 指示,命令,规定,要求

die 锻模,冲模

die casting 压力铸造,金属型铸造,压铸件

dielectric [daii'lektrik] a. 不导电的,绝缘的,电介质的

dielectric constant 介电常数,介质常数

difficult-to-machine material 难加工材料

dilemma [di'lemə] n. 困境,进退两难

dimensioning 标注尺寸

direct costs and indirect costs 直接成本与间接成本

direct current 直流电
disassemble [disə'sembl] v. 拆卸,拆除,分解
discipline ['disiplin] n. 学科,科目
disclose [dis'kləuz] v. 揭露,揭发,揭开,泄露
discount ['diskaunt] n. 折扣 v. 打折
discrepancy [dis'krepənsi] n. 不同,不符合,矛盾,差异,分歧
disintegrate [dis'intigreit] v. (使)分散[离,解,化],粉碎,切碎
dislocation [dislə'keiʃən] n. (晶体格子中)位移,位错
dismount [dis'maunt] n. ;v. 卸下,拆除,拆卸
displaced 代替的,替换的
disposal [dis'pəuzəl] n. 废弃物清除,处置,处理
dissipation [disi'peiʃən] n. 消散,扩散,散失,消耗
distance education 远程教育
distortion [dis'tɔ:ʃən] n. 变形,扭曲,畸变
distribute [dis'tribju:t] v. 分配
distributed intelligence 分布式智能
disturb [dis'tə:b] v. 扰动,干扰,妨碍
disturbing force 干扰力
disulfide [dai'sʌlfaid] n. 二硫化物
diversification [dai,və:sifi'keiʃən] n. 多样化,变化
diversity [dai'və:siti] n. 不同,异样性,多种多样性
divider [di'vaidə] n. 分规,两脚规
digitize ['didʒitaiz] v. 数字化
DNC (direct numerical control) 直接数字控制
domain [də'mein] n. 范围,领域
dominate ['dɔmineit] v. 支配,占优势
double-counted 被重复计算
download 下载
downtime ['dauntaim] n. 停机时间,发生故障时间,修理时间
draft angle n. 拔模斜度

drafter [drɑːʃtə] n. 制图者,描图者,制图机械
drafting ['drɑːftiŋ] n. 制图,起草
drafting board 绘图板
draw conclusion 得出结论
dress 对砂轮进行修锐和整形
dresser 修整器(为了获得一定的几何形状和锐利磨削刃,用金刚石等工具对砂轮表面进行修整的装置)
dressing ['dresiŋ] n. (砂轮)修整(产生锋锐的磨削刃)
drill press 钻床
drilling ['driliŋ] n. 钻孔
drive line 动力传动系统,传动轴装置
drive mechanism 驱动机构,驱动装置
drive motor 驱动电机
drive train 动力传动系统
driver [draivə] n. 驱动装置
driver motor 主驱动电动机
DSP = digital signal processor 数字信号处理器
duplicating lathe 仿形车床
duplication [djuːpliˈkeiʃən] n. 复制,副本
durable ['djuərəbl] a. 耐用的,耐久的;n. 耐用的物品
DXF (data exchange format) 数据交换格式
dynamic performance 动态特性,动力特性
dynamics [daiˈnæmiks] n. 动力学

E

EDM 电加工
effectiveness [iˈfektivnis] a. 效率,效能,效果
effortless [ˈefətlis] a. 容易的,不费力气的
elaborate [iˈlæbərit] v. 详细描述,详细阐述
elasticity [elæsˈtisiti] n. 弹性(变形),伸缩性,弹性力学

elastomer [i'læstəmə] n. 弹性体,合成橡胶

elective course 选修课程

electric drive 电气传动,又称"电力拖动"(生产过程中,以电动机作为原动机来带动生产机械,并按所给定的规律运动的电气设备)。

electrochemical [i'lektrəu'kemikl] a. 电化学的

electro-mechanical [i'lektrəumi'kænikəl] a. 机电的,电动机械的

eligible ['elidʒəbl] a. 符合被推选条件的,合格的

embark [im'bɑ:k] v. 从事,着手,开始

embed [im'bed] v. 嵌入,埋入,植入

embodiment [im'bɔdimənt] n. 具体化,具体表现,具体装置

emerging [i'mə:dʒiŋ] a. 新兴的,新出现的

emission [i'miʃən] n. 发射,放出,排出物

employee involvement 雇员参与

empower [im'pauə] v. 授权给,准许,授予...的权利(资格)

empowerment [im'pauəmənt] n. 授予权利

emulate ['emjuleit] v. 仿真,模仿,模拟

en mass 全部,全部地,整个地

enclosure [in'kləuʒə] n. 外壳,机壳,罩

encoder [in'kəudə] n. 编码器(一种能提供位置反馈和速度反馈的测量装置)

encompass [in'kʌmpəs] v. 环绕,包围,包括,包含

encounter [in'kauntə] v. ;n. 遇到,遇见,遭遇

encryption [in'kripʃən] n. 加密,编密码

end effector 末端作用器,末端器

end mill 立铣刀

end product 最后结果,最终产品,成品

end user 终端用户,最终用户

energize ['enədʒaiz] v. 激发,给予...电压,使...带电

engage [in'geidʒ] v. 使从事于,参加

engine lathe 普通车床

engineered *a.* 设计的

engineering constraint 工程技术约束(限制)

engineering time 维修时间,工程维护时间,维护检修时间

enhanced 增强的,增强型的

enormous [i'nɔ:məs] *a.* 巨大的

entrench [in'trentʃ] *v.* 牢固树立,确立,确定

entry-level 入门水平的

enumerate [i'nju:məreit] *v.* 数,列举,计算

epitome [i'pitəmi] *n.* 梗概,概括,缩影,集中表现

equation [i'kweiʃən] *n.* 方程式,公式

equilibrium [i:kwi'libriəm] *n.* 平衡,均衡

equivalent [i'kwivələnt] *a.* 相当的,相等的;*n.* 同等物,相等物

ergonomics [ˌə:gəu'nɔmiks] *n.* 人机工程学,人与机械控制

excessive [ik'sesiv] *a.* 过度的,极度的,非常的

exclusive [iks'klu:siv] *a.* 排它的,排外的,专用的,独占的

execute ['eksikju:t] *v.* 实行,完成,实现,实施

executive director 执行理事,常务董事

expand and contract 膨胀与收缩

expanding [iks'pændiŋ] *n.* 扩充,扩展,扩张

expansion coefficient 膨胀系数

expediency [ik'spi:diənsi] *n.* 方便,便利,权宜之计

experimental study 实验研究,实验

expertise [ekspə'ti:z] *n.* 专门知识,专长,专家鉴定

exponential [ekspəu'nenʃəl] *a.* 指数的,幂的;*n.* 指数

exposed wire 露在外面的电线,明线

expound [ik'spaund] *v.* 详细说明,解释

expressed warranty 明确保证

extraneous [eks'treinjəs] *a.* 外部的,附加的,无关的,不重要的

extreme [ik'stri:m] *n.* ;*a.* 极端(的),极度(的);*n.* 极端的事物

F

facilitate [fə'siliteit] v. 推动,促进,帮助
facilitator 协调员
facility [fə'siliti] n. 工厂,设备
facing ['feisiŋ] n. 端面车削
faculty ['fæklti] n. (大学的)院、系,全体教师
faculty grant 教师科研基金
family of part 零件族
far-flung 分布广的,范围广的,遥远的,漫长的
fastener [fɑ:snə] n. 紧固件,联接件
fat [fæt] n. 脂肪,油脂;a. 油脂的,多脂的
fatigue [fə'ti:g] n. ;v. 疲劳
fault-tolerant computer 容错计算机
feasibility [fi:zə'biliti] n. 可行性,可能性,现实性
feasible ['fi:zəbl] a. 可行的,行得通的,合理的
feed [fi:d] (fed,fed) v. ;n. 进给,进给量
feed rate 进给速度
feedback ['fi:dbæk] n. 反馈,反应
feedback loop 反馈回路
fiber-optic light guide 光导纤维
field service 现场维修
fillet ['filit] n. 圆角,倒角
finalize ['fainəlaiz] v. (把计划,稿件等)最后确定下来,定案,完成工作
finding ['faindiŋ] n. 研究结果,试验结果,观察结果
fine adjustment 精密调整
fine-tuning 微调,细调
finish turning 精车
finite element analysis (FEA) 有限元分析
firewall 防火墙

first-line supervisor 第一线的管理人员
fitness for use 适用性(即产品或服务能够成功地符合用户需要的程度)
fitting 安装,装配
fixed cost 固定成本
fixed expense 固定费用
fixed-sequence robot 固定顺序机器人,固定程序机器人
fixture ['fikstʃə] n. 夹具,设备,装置
fixture base 夹具体
fixture crib 夹具室
flange [flæŋdʒ] n. 边缘,凸缘,法兰
flawless ['flɔːlis] a. 无瑕的,无缺点的,完美的
flexible coupling 弹性(挠性)联轴器
flexible machining cell 柔性加工单元
flexible manufacturing cell (FMC) 柔性制造单元
floodgate ['flʌdgeit] n. 泄洪闸门,水闸,大量
floor space 占地面积,房屋面积
fluctuate ['flʌktjueit] v. 脉动,振荡,变化
fluoride ['fluəraid] n. 氟化物
flywheel ['flaihwiːl] n. 飞轮,惯性轮
FMEA = failure mode and effect analysis 故障模式和效应分析
FMS (flexible manufacturing system) 柔性制造系统
foolproof ['fuːlpruːf] a. 十分简单,不会出错的
foreign material 外来材料,异物,杂质
foreign-built machine tool 外国制造的机床
forging ['fɔːdʒiŋ] n.;a. 锻造(的),模锻,锻件
fork truck 叉式起重车
forklift truck 叉车,叉式升降装卸车
form grinding 成形磨削
forming die 成形钢模,锻模
formulation [fɔːmjuˈleiʃən] n. 列方程式,列出公式,(有系统的)阐述

foundation [faun'deiʃən] n. 基础,地基,底座,机座
fraction ['frækʃən] n. 分数,比值,几分之一
fractional ['frækʃənl] a. 分数的,小数的
fracture ['fræktʃə] v. ;n. 破碎,断裂
fracture toughness 断裂韧度(材料抵抗裂纹启裂和扩展的能力)
fragile ['frædʒail] a. 易碎的,易毁坏的
framework ['freimwə:k] n. 构架,框架,结构
freehand ['fri:hænd] a. 徒手画的
friction ['frikʃən] n. 摩擦,摩擦力
fringe [frindʒ] n. 边缘;a. 边缘的,附加的,较次要的
fringe benefit 附加福利,补贴,福利金
front panel 面板
frontend 前端
full-scale 与原物大小一样的,全面的,全部的
functionality [fʌŋkʃən'æliti] a. 功能性
fundamental [fʌndə'mənt] n. 基本原理;a. 基本的
furnish ['fə:niʃ] v. 提供,供给
fuzzy logic 模糊逻辑

G

gage=gauge [geidʒ] n. 测量仪表,传感器,检验仪器
gear [giə] n. 齿轮
General Motors Corporation (美国)通用汽车公司
general-purpose 多方面的,多种用途的
generative ['dʒenərətiv] a. 能生产的,创成的
generative approach 创成法
geographic [dʒiə'græfik] a. 地理上的,地区性的
geometric modeling 几何建模,几何模型建立,形状模型化
geometric primitive 几何图元
government regulation 政府法规

graphics ['græfiks] n. 制图学,图形学,图解
graphite ['græfait] n. 石墨
grapple with a problem 设法解决问题
grease [gri:s] n. 润滑脂,黄油
grinder 磨床
grinding ['graindiŋ] n. 磨削;a. 磨削的
grinding quill 磨床主轴
groove [gru:v] n. 凹槽,沟;v. 开槽于
ground part 经过磨削加工的零件
ground rule 程序
groundwork ['graundwə:k] n. 基础工作,根据
group technology 成组技术
grouping ['gru:piŋ] n. 分组,分类,归类
guarantee [gærən'ti:] n. ;v. 保证(书,人),担保,承诺
guard n. 保护(器,装置),防护(器,罩,装置)
guideline 方针,准则,指导方针

H

hamper ['hæmpə] v. 妨碍,阻碍,阻止
haphazard ['hæp'hæzəd] n. 偶然性,任意性;a. 杂乱的,任意的
hard [hɑ:d] a. 硬的,淬硬的
hard-part machining(HPM) 硬态切削
harden ['hɑ:dn] v. 淬火,增加硬度
hardness ['hɑ:dnis] n. 硬度
hardwired a. 电路的,硬件实现的,硬连线的
harness ['hɑ:nis] v. 利用,治理
have as 把…当作
hazardous ['hæzədəs] a. 危险的
headstock ['hedstɔk] n. 头架,主轴箱
heat check 热裂纹,热致裂纹

heat exchanger 热交换器
heat-treated 热处理的,热处理过的
heavy-duty 重型的,重载的,耐用的,高功率的
heavy-handed a. 严厉的,不明智的,苛刻的
helical spring lockwasher 弹簧垫圈
helicopter rotor blade 直升飞机转子叶片
hertz [hə:ts] n. 赫兹(频率单位)
Hertzian contact stress 赫兹接触应力
hexagonal [hek'sægənəl] a. 六角形的,六边形的
hierarchical [ˌhaiə'rɑ:kikəl] a. 分层的,层次的,递阶的
hierarchical control 递阶控制,层次控制
hierarchy ['haiərɑ:ki] n. 体系,分层,层次
high power density 大功率密度
high volume production 大批量生产
high-powered a. 大功率的,高性能的
high-resolution graphics 高分辨率图形
high-tech 高科技,高技术的
hinge [hiŋdʒ] n. 铰链,装铰链
Hitachi [hi'tɑ:tʃi] n. 日立
holder 夹持装置,固定器
home page 主页
homework 预先的准备工作
horizontal machining center 卧式加工中心
hose down 用水龙带冲洗,用软管洗涤
hot isostatically pressed 热等静压的
hot pressed 热压的
hot-pressed alumina insert 热压氧化铝刀片
housekeeping 辅助工作,服务性工作
housing ['hauziŋ] n. 壳,套,罩,盒,箱
hub [hʌb] n. 轮毂

hybrid ['haibrid] n. 混合物;a. 混合式的
hydraulic [hai'drɔ:lik] a. 水力的,液力的,液压的
hydraulic cylinder 液压缸
hydraulic motor 液压马达
hydraulically actuated 液压驱动的
hydraulics 水力学,液压系统
hydrodynamical [haidrəudai'næmikəl] a. 流体动力(学)的
hydrostatic bearing 流体静压轴承
hydrostatical [haidrəu'stætikəl] a. 流体静力(学)的,液压静力的
hypothesis [hai'pɔθisis] (pl. hypotheses) n. 假说,假设,前提

I

IC = integrated circuit 集成电路
icing ['aisiŋ] n. (糕饼表层的)糖霜,酥皮
icon ['aikɔn] n. 图标,肖像
identify [ai'dentifai] v. 识别,确定,发现
IEEE 电气和电子工程师协会
IGES (initial graphic exchange specification) 初始图形交换规范
illustrate ['iləstreit] v. 举例说明,阐明,图解
illustrative ['iləstritiv] a. 说明的,解说的,例证的
imbalance [im'bæləns] n. 不平衡,不均衡
impel [im'pel] v. 推进,推动,驱使,促成,刺激
impeller [im'pelə] n. 叶轮,涡轮,推进器
imperfect [im'pə:fiktli] ad. 有缺点地,不完善地
impetus ['impitəs] n. 原动力,促进,推动
implementation [implimen'teiʃən] n. 履行,实施,实现,执行过程
implication [impli'keiʃən] n. 关系,隐含,意义,本质,实质
implied warranty 内在保证
impurity [im'pjuəriti] n. 杂质,夹杂物
inaccessible [inæk'sesəbl] a. 不能接近的,不能进入的

inaccuracy [in'ækjurəsi] n. 误差,不准确
Inc. [ink] (=incorporation) n. 公司,团体,法人
inception [in'sepʃən] n. 起初,开端,开始
incipient [in'sipiənt] a. 开始的,起首的,初步的
incline [in'klain] n. ;v. 倾向,倾斜
Inconel 因科合金,铬镍铁耐热耐腐蚀合金
incontrovertible [inkɔntrə'və:təbl] a. 无可争辩的,反驳不了的,颠扑不破的
incorporated 合为一体的
incur [in'kə:] v. 招致,承担,遭受
independent variable 自变量
index ['indeks] v. 转换角度,转位,换挡,分度,索引,检索,目录
individual island of automation 独立自动化制造岛
induction [in'dʌkʃən] n. 感应
induction motor 感应电机
industrial economy 工业经济
infant mortality 早期失效率
infeasible [in'fi:zəbl] a. 不能实行的,办不到的,不可能的
infeed 横向进磨,切入磨法,横切(进给)
inference ['infərəns] n. 推论,结论,含意
infiltrate ['infiltreit] v. 渗透,渗进,过滤;n. 渗入物
information age 信息时代
information highway 信息高速公路
inherently [in'hiərəntli] ad. 固有地,本征地,内在地
inhouse (机构)内部的,自身的
initiate [i'niʃieit] v. 开始,创始,起始,着手,引起;n. 首创精神,主动,积极性
injection molding 注射成型,注塑成型
injection-molded 注射成型的,注塑成型的
insert 嵌入件,埋入件,刀片

insistence [in'sistəns] n. 坚持,坚决主张
instability [instə'biliti] n. 不稳定性
installation [ˌinstɔː'leiʃən] n. 整套装置,设备,结构,安装
instructional [in'strʌkʃənəl] a. 教学的,教育的
instructor 任课教师
instrumentation [instrumen'teiʃən] n. 仪器,工具
insulation [insju'leiʃən] n. 绝缘,隔热,绝缘体
integral sign 积分号
integrated ['intigreitid] a. 综合的,完整的
integrated approach 综合方法
integrated control system 综合控制系统
integration [inti'greiʃən] n. 集成,综合
integrator 综合者,合成者
intelligent robotics 智能机器人技术
interactive [intər'æktiv] a. 交互式的,人机对话的
interactive graphics-based 基于交互式图形的
interactive instruction 交互式教学
interchangeability ['intəˌtʃeindʒə'biliti] a. 互换性,可替换性
interchangeable [intə'tʃeindʒbl] a. 可互换的,可拆卸的,通用的
interface [intə'feis] n. 界面,相互作用面,接口
interfere [intə'fiə] v. 干涉,干扰,同...抵触,冲突(with)
interfere with 妨碍,干涉,与...抵触
intermediate shaft 中间轴
intermittent [intə'mitənt] a. 间歇的,断续的,周期性的
International Institution for Production Engineering Research(CIRP)
国际生产工程学会
internship n. 实习生,实习
interpolation [intəːpəu'leiʃən] n. 插值,插入
interpretation [intəːpriː'teiʃən] n. 解释,说明
intersection [intə'sekʃən] n. 相交,十字路口,交叉点

intersperse [intə'spə:s] v. 散布,散置
Intranet ['intrə'net] n. 企业内部互联网
introduction 引言,导论,前言
introductory [intrə'dʌktəri] a. 引言的,导言的
inventory ['invəntri] n. 清单,报表,库存量
inventory control 库存管理
inverse correlation 逆相关,反相关
in-house [in'haus] a. 国内的,机构内部的,公司内部的
in-person 亲身的,亲自的,在现场的
in-process a. (加工,处理)过程中的
in-process sensing 加工过程中的检测
irregular [i'regjulə] a. 不规则的
irrelevant [i'relivənt] a. 不相关的,没关系的
irresponsible [iris'pɔnsəbl] a. 不负责任的,不可靠的
isolation [aisə'leiʃən] n. 孤立,隔离
italicize [i'tælisaiz] v. 用斜体字印刷
iteration [itə'reiʃən] n. 反复,重复,迭代
iterative ['itərətiv] a. 反复的,迭代的

J

jet turbine engine 喷气涡轮发动机
jig grinder 坐标磨床
jigs and fixtures 夹具(用以装夹工件和引导刀具的装置)
joining ['dʒɔiniŋ] n. 连接,连接物
joint [dʒɔint] n. ;v. 结合,连接
joint venture 合资,合资企业
journal 轴颈
journal bearing 滑动轴承
judicial [dʒu:'diʃəl] a. 司法的,法院的,公正的,明断的
junk [dʒʌŋk] n. 废物,废品;v. 把...(当作废物)丢弃

just in time production 准时生产(即只在需要的时候,按需要的数量生产所需的产品)

justify ['dʒʌstifai] v. 证明...是正确的,认为...有理由

K

key parts supplier 关键零件供货厂商

keyway 键槽

keyway cutter 键槽铣刀

kinetic [kai'netik] a. 运动的,动力的,活动的

kinetic energy 动能

knowledge base 知识库(专家系统的一部分,包括解决某一专门领域问题所需的细节和规则)

knowledge engineer 知识工程师,建立专家系统的编程人员

knowledge-based 基于知识的

know-how ['nəuhau] n. 专门技能,窍门,诀窍,生产经验

L

lab 实验室

labor-intensive 劳动密集型的

large-scale 大规模的,大范围的

lathe [leið] n. 车床;v. 用车床加工

lattice ['lætis] n. 晶格

lead [led] n. 铅,铅制品

lead screw 导杆,丝杠

lead time 订货至交货之间的时间,研制周期,交付周期

leading [li:diŋ] a. 主要的,最重要的

lead-through teaching 示教的,仿效的

lecture hall 大教室,讲演厅

left hand thread 左旋螺纹

lessen ['lesn] v. 减少,缩小,减轻

lesser developed country 欠发达国家,不发达国家
liaison [li'eizən] n.;v. 联络,联系(人);协作(with)
life cycle 寿命周期,耐用周期
like magnetic pole 相同的磁极
likelihood ['laiklihud] n. 可能,可能性
limit switch 限位开关,极限开关
line shaft 动力轴,主传动轴
linear ['liniə] a. 线的,直线的,线性的
linear bearing 直线轴承
linear motor 直线电动机
linkage ['liŋkidʒ] n. 连接,联系,连杆机构
list cost 定价,价目表所列之价格,标价
literature search 文献检索
litigation [liti'geiʃən] n. 诉讼,打官司
load line 负荷线,负载线,载荷曲线
loading condition 载荷条件,负荷条件
load-bearing capability 承载能力
lobe [ləub] n. 凸起,凸起部
lobing ['ləubiŋ] n. (圆柱的)凸角
local area network 局域网,本地网
log [lɔg] n. (工程、试验等的)工作记录
longitudinal [lɔŋdʒi'tjuːdinəl] a. 纵向的,长度的,轴向的
loom [luːm] n. 织布机
lower deviation 下偏差
lower-case letter 小写字母
lubricant ['ljuːbrikənt] n. 润滑剂,润滑材料
lubrication [luːbri'keiʃən] n. 润滑
lubricity [luː'brisiti] n. 润滑,润滑能力
lump [lʌmp] v. 把...合在一起,总结,概括

M

Mach [mɑːk] n. 马赫(速度单位),速度与音速之比
machine configuration 机器配置
machine function 机器功能,机器作用
machine tool 机床
machining [məˈʃiːniŋ] n. 机械加工,切削加工
machining center 加工中心
machinist [məˈʃiːnist] n. 机械工人,机工
made to order 定制的,定做的
magnetize [ˈmægnitaiz] v. 使磁化
main frame computer 大型计算机
mainframe 主计算机,大型计算机
mainshaft 主轴
maintainable 可维修的,可保养的
maintenance [ˈmeintinəs] n. 维持,(技术)保养,维修
major diameter (螺纹)大径,外径
manipulator [məˈnipjuleitə] n. 操作器,操作装置,机械手,操作机
manufacturability 可制造性,工艺性
manufacturing [mænjuˈfæktʃəriŋ] a. 制造的,制造业的; n. 制造,生产
manufacturing cell 制造单元,加工单元
manufacturing process 制造过程,制造方法
manufacturing step 工步
manuscript [ˈmænjuskript] n. 稿件
mapping 映像,映射,变换,交换
mar [mɑː] v. 损坏,破坏
marginal [ˈmɑːdʒinl] a. 临界的,勉强合格的
marketing department 销售部门
marketing division 销售部门
marketplace [ˈmɑːkitˈpleis] n. 市场

mass production 大规模生产

master production schedule 总生产进度表

material requirement planning 物料需求计划(制造企业以在指定日期生产出指定产品为目标,确定生产所需的原料采购和部件装配的信息系统)

material specification 材料规格,材料明细表

material-handling 物料输送

mating part 配合件,配合部分

matrix ['meitriks] (pl. matrices 或 matrixes) n. 基体,母体,本体

mean 平均值

mean time between failures 平均故障间隔时间

means ['mi:nz] n. 手段,方法;v. 意味,想要

means of communication 交流工具(手段),通讯工具(手段)

mechanical advantage 机械效益,机械增益

mechanical transmission 机械传动

mechanism ['mekənizəm] n. 机构,机械装置

mechatronic 机电一体化的

memorandum [memə'rændəm] n. 备忘录,便笺,便函

merit ['merit] n. 优点,特征,价值;v. 值得,有...价值

metal removal 金属切削

metallic [mi'tælic] a. 金属的

metallurgy [me'tælədʒi] n. 冶金学,冶金

metalworking ['metlwə:kiŋ] n. 金属加工,金工

methodology [meθə'dɔlədʒi] n. 方法(学,论),分类法

metric thread 公制螺纹

metrology [mi'trɔlədʒi] n. 计量学,测量学,计量制

micro ['maikrəu] n.;a. 微型(的),微细(的),微观(的)

microcontroller [maikrəukən'trəulə] n. 微控制器

micromanufacturing 微细加工,微细制造

milling ['miliŋ] n. 铣削

minimize ['minimaiz] v. 使...成为最小,最小化

misalignment ['misəlainmənt] n. 未对准,轴线不重合,安装误差
misspelling [mis'speliŋ] n. 拼写错误
misuse ['mis'ju:z] v. ; n. 错用,误用,滥用
mitigate ['mitigeit] v. 使缓和,减轻,防止
modeling ['mɔdliŋ] n. 建模,造型
modest ['mɔdist] a. 适度的,合适的
modular ['mɔdjulə] a. 制成标准组件的,预制的,组合的
modular fixturing 组合夹具
modular system 模块化系统
modularity [mɔdju'læriti] n. 模块性,模块化,调制性
module ['mɔdju:l] n. 模数,模件,组件,可互换标准件
modulus ['mɔdjuləs] (pl. moduli) n. 模数,模量,系数
modulus of elasticity 弹性模量
mold 模具,铸模
molded part 模制零件
molybdenum [mɔ'libdinəm] n. 钼
Monel [məu'nel] n. 蒙乃尔铜镍合金
monitor ['mɔnitə] n. 监视,监测,监督
monolithic [mɔnəu'liθik] a. 单一的,统一的,整体的
monolithic system 单片系统
monotonic [mɔnə'tɔnik] a. 单调的
mortality [mɔ:'tæliti] n. 致命性,失败率,死亡率
MOSFET 金属氧化物半导体场效应晶体管
motivation [məuti'veiʃən] n. 刺激,激发,诱导,动机
motive ['məutiv] n. 动机,目的
motive power 动力,原动力
motoring ['məutəriŋ] n. 驾驶汽车;a. 汽车的
motto ['mɔtəu] n. 座右铭,格言,题词,标语
mount [maunt] n. 固定,固定件;v. 安装,固定
mounting ['mauntiŋ] n. 安装,安置,装配,配件,固定件

multidisciplinary [mʌlti'disiplinəri] a. 包括各种学科的,多种不同学科的

multifaceted a. 多方面的,多层次的

multifunctional a. 多功能的

multifunctional or interdisciplinary teams 多功能或多学科小组

multilayer ['mʌltileiə] n. 多层;a. 多层的

multiple ['mʌltipl] a. 多样的;n. 倍数,若干 v. 成倍增加

Murphy's law 墨菲法则(一种幽默的规则,它认为任何可能出错的事终将出错)

mutual respect 相互尊重

mutually exclusive 互斥的,不相容的

N

natural language 自然语言(人类写或说的语言)

navigate ['nævigeit] v. 使通过,沿...航行

NC package 数控程序包

net production 净生产量

neutralize ['nju:trəlaiz] v. 使中和,使中立

niche market 瞄准机会的市场(指专门瞄准机会,做因市场不大而别人不做的产品,从而获得丰厚的利润)

nickel ['nikl] n. 镍

node [nəud] n. 节点,交点,中心点

nomenclature [nəu'menklətʃə] a. 名词,术语,术语表

nominal ['nɔminl] a. 标定的,额定的,极小的,按计划进行的;n. 标称,额定

nonconformity ['nɔnkən'fɔ:miti] n. 不适合,不一致

nonmonetary [nɔn'mʌnitəri] a. 非货币的

nonproductive [nɔnprə'dʌktiv] a. 非常生产性的,与生产无直接关系的

nontechnical [nɔn'teknikl] a. 非技术性的

normal distribution 正态分布

normal curve 正态曲线
normal force 法向力
not to scale 不按比例
notation [nəu'teiʃən] n. 符号,记号,注释
noticeable ['nəutisəbl] a. 引人注意的,显著的
notification [nəutifi'keiʃən] n. 通知,告示,布告
nut [nʌt] n. 螺帽,螺母

O

objectivity [ɔbdʒek'tiviti] n. 客观性,客观现实
OD cylindrical grinder 外圆磨床
offshore ['ɔ:fʃɔ:] a. 海外的,国外的,在国外建立的
off-the-shelf 成品的,畅销的,现成的,流行的
oil in 进油,进油口
oil out 出油,出油口
on the verge of 接近于,濒临于
one's strong suit 优点,长处
one-off 一次性的,单件的;一次性事物,单件生产
one-size-fits-all 全能的,可以符合或适用各种要求的
online ['ɔnlain] a. 联机的,在线的
online web-based learning 基于网络的在线学习
op amps = operational amplifier 运算放大器
open-ended a. 可扩展的,无终止的,能适应未来发展的,未确定的
operating condition 运行条件,工作状况,操作条件
operating cycle 工作循环,操作循环
operating procedure 操作程序,工作方法
operating temperature 工作温度
operation 工序
operation sheet 工序卡片
optical goods 光学产品

optimally ['ɔptiməli] ad. 最佳地,最优地,最恰当的,最适宜地
optimum ['ɔptiməm] a. ;n. 最佳(的,值,状态,条件,方式),最适宜,最有利的
orient ['ɔ:riənt] v. 确定方向
orientation [ɔ:rien'teiʃən] n. 介绍,针对新形势的介绍性指导
original design 原创性设计,创新设计
orthographic [ˌɔ:θə'græfik] a. 直角的,正投影的
orthographic projection 正投影
oscillate ['ɔsileit] v. (来回)摆动,振荡,摇摆
oscillation [ɔsi'leiʃən] n. 振荡,摆动
out of round 不(很)圆,失圆
outline ['autlain] v. 概括地论述
outperform [autpə'fɔ:m] v. 做得比...好,胜过
outrageously [aut'reidʒəsli] ad. 令人震惊地,令人不能容忍地
outweigh [aut'wei] v. 优于,多于,比...重要,胜过
out-of-tolerance 公差范围以外的
overhead ['əuvəhed] n. 经常费,管理费,杂费
overhead hoist 提升机,桥式起重机
overlap [əuvə'læp] v. ;n. 重叠,叠加,相交,交错
override [əuvə'raid] v. 超过,克服
overstress [əuvə'stres] n. ;v. 过载,超载,过度应力
overview ['əuvəvju:] n. 总的看法,概观,综述
overwhelming [əuvə'hwelmiŋ] a. 压倒的,不可抵抗的,优势的
oxide ['ɔksaid] n. ;a. 氧化物(皮,层,的)

P

package ['pækidʒ] n. ;v. 包装,捆,束,程序包
packaging ['pækidʒiŋ] n. 打包,装箱,包装
pallet ['pælit] n. 随行夹具,随行托板
palletizing ['pælitaiz] v. 码垛堆积,放在托板上

panacea [pænə'siə] n. 灵丹妙药,解决一切问题的方法
parametric [pærə'metrik] a. 参数的
parenthesis [pə'renθisis] n. (pl. parentheses) 插句,括弧,圆括号
parenthetically [pærən'θetikəli] ad. 顺便地,作为插句
part family 零件族,零件组
part list 零件明细表,材料清单
part number 零件序号
parting ['pɑ:tiŋ] a. 分离的,离别的;n. 分离,切断
partition [pɑ:'titʃən] n.;v. 划分,区分,分割,分类
party (诉讼,协约,会议等)一方,当事人
passageway ['pæsidʒ'wei] n. 通道,通路
pattern ['pætən] n. 模,木模,铸模
pedagogy ['pedəgɔgi] n. 教学法,教育学
penetration [peni'treiʃən] n. 穿透,穿透能力,穿透深度
per se 自身,本来,性(本)质上
perceptible [pə'septibl] a. 可以感觉得到的,看得出的,显而易见的
perimeter [pə'rimitə] n. 周边
peripheral [pə'rifərəl] a. 周边的,外围的;n. 外部设备,附加设备
periphery [pə'rifəri] n. 周边,圆周,圆柱(体)表面
permeate ['pə:mieit] v. 渗入,渗透,弥漫,充满
permutation [pə:mju'teiʃən] n. 变更,置换,重新配置,排列
perpendicular [pə:pən'dikjulə] a. 与...垂直的;n. 垂直,垂线
perpetuate [pə'petjueit] v. 使永存,保全,维持
pertinent ['pə:tinənt] a. 恰当的,贴切的,中肯的,与...有关的(to)
phase [feiz] n. 相,发展阶段;v. 使分阶段(按计划)进行,逐步采用
pictorial [pik'tɔ:riəl] a. 用图表示的,图解的
PID 比例积分微分
pilot ['pailət] a. 引导的,导向的,(小规模)试验性的,试点的
pitch [pitʃ] n. 螺距
pitch diameter 螺纹中径

pitfall ['pitʃɔ:l] n. 缺陷,易犯的错误,隐患
pivotal ['pivətl] a. 枢轴的,关键性的,非常重要的
placement ['pleismənt] n. 放置,布置
planer ['pleinə] n. (龙门)刨床
plasmas ['plæzmə] n. 等离子
plastic injection molding 注塑成型
plastic processing machine 塑料加工机械
platinum ['plætinəm] n. 铂,白金
playback robot 示教再现式机器人,重现机器人
plotter ['plɔtə] n. 绘图仪
point-to-point 点对点,点位
polarity [pəu'læriti] n. 极性
polycrystalline [pɔli'kristəlain] a. 多晶的
polymer ['pɔlimə] n. 聚合物,高分子材料
polymeric [pɔli'merik] a. 聚合物的,高分子材料的
positioner [pə'ziʃənə] n. 定位元件,定位装置
positioning [pə'ziʃəniŋ] n. 定位,位置控制
post processing 后处理,后加工
potted ['pɔtid] a. 封装的,防水包装的
powder metallurgy 粉末冶金
power conditioner 动力(功率)调节器
power dissipation 功率耗散,功率消耗
power tool 电动工具
prearrange ['pri:ə'reindʒ] v. 预先安排,预定
preconceive [pri:kən'si:v] v. 预想,事先想,事先做出的
precondition ['pri:kən'diʃən] n. 前提,先决条件
predominant [pri'dɔminənt] a. 占主导地位的,(在数量上)占优势的
preempt [pri'empt] v. 优先购买,先取,先占
preheat [pri:'hi:t] n. 预先加热
premade 预先做的

premature [preməˈtjuə] a. 过早的,未成熟的,不到期的
premise [ˈpremis] n. 前提,前言,根据
preplan [priːˈplæn] v. 规划,预先计划
preproduction [priːprəˈdʌkʃən] a. 生产前的,试制的,试生产的;n. 试生产
prerequisite [priːˈrekwizit] n. 先决条件 a. 首要必备的
presentation 讲述,展示,介绍
preservative [priˈzəːvətiv] a. 保存的,防腐的;n. 防腐剂,保存剂
pressurize [ˈpreʃəraiz] v. 增压,对...加压,产生压力
pretest [ˈpriːtest] n.;v. 事先试验,预先测试
prevalent [ˈprevələnt] a. 流行的,盛行的,普遍的
prevention cost 预防费用
preview [ˈpriːˈvjuː] n.;v. 预览,事先查看,预演
prime mover 原动力,原动机(转化自然能使其做功的机器或装置)
primitive [ˈprimitiv] a. 原始的,基本的;n. 基元,图元
printed-circuit board 印刷电路板
prism [ˈprizm] n. 棱柱
probability [prɔbəˈbiliti] n. 概率,可能性,可能发生的事
probe [prəub] n. 探针,探头,探测器
procedurally [prəˈsiːdʒərəli] ad. 程序地,程序性地
process capability 工序能力,设备加工能力
process capability index 工序能力指数
process control 过程控制,工艺管理
process plan 工艺规程,生产工艺设计
process planning 工艺过程设计(编制各种工艺文件和设计工艺装备等过程)
process route 工艺路线
process sheet 工艺过程卡,工艺卡
process variation 加工偏差,加工变化范围
procurement [prəˈkjuəmənt] n. 获得,采购,供货合同
product liability 产品责任

product specification 产品规格,产品技术规范
product use 产品用途
production facility 生产设备
production line 生产线,流水线
production process 生产过程
production schedule 生产进度计划
productivity [prɔdʌk'tiviti] n. 生产力,生产率
professional [prə'feʃənl] n. 特性,专长,专业,专业人员,内行
profitable 有利可图的
progress report 进度报告,进展报告
prompt [prɔmpt] n. 提示
promptly ['prɔmptli] ad. 迅速地
propeller [prə'pelə] n. 螺旋桨,推进器
proprietary [prə'praiətəri] a. 专利的,专有的,独占的
pros and cons 正反面,优缺点,正反两方面的理由
prototype ['prəutətaip] n. 原型,样机,样品
proven a. 被证实的,可靠的
provision [prə'viʒən] n. (预防)措施,保证,保障
proximate ['prɔksimət] a. 最近的,直接的
proximate cause 直接原因
proximity [prɔk'simiti] n. 接近,贴近,近程
proximity sensor 接近传感器
PSI = pounds per square inch 磅/平方英寸
publications 出版物,发行物
publicity [pʌb'lisiti] n. 宣传材料,广告
pulley ['puli] n. 滑轮,皮带轮
pulse [pʌls] n. 脉冲
punched card 穿孔卡片
purchase order 购货订单
purpose-built 为特定目的而建造的,特别的,专用的

Q

QC (quality control) 质量控制

quadrant ['kwɔdrənt] n. 象限,四分之一圆

qualitative ['kwɔlitətiv] a. 定性的,性质上的

quality assurance 质量保证

quality improvement 质量改进

quality of conformance 符合质量标准的程度,符合质量(用设计质量和制造质量,或预期质量和实际质量的差值来评价产品制造质量的水平)

quality program 质量保证计划,质量规划

quality system 质量体系,质量系统

quantitative ['kwɔntitətiv] a. 数量的,定量的

quantify ['kwɔntifai] v. 确定数量,量化

R

rack [ræk] n. 架,支架,齿条,滑轨

radial ['reidiəl] a. 径向的;n. 径向

radially ['reidiəli] ad. 径向地

radically ['rædikəli] ad. 根本地,安全地,主要地

radius ['reidjəs] n. (pl. radii ['reidiai]) 半径,半径范围

random ['rændəm] a. 随机的

random order 随机顺序,任意顺序

randomly ['rændəmli] ad. 任意地,偶然地

rapport [ræ'pɔːt] n. 关系,联系

reactive [riːæktiv] a. 反应的,易起反应的,对反应敏感的

ready-made 现成的,做好的

real life 真实的,实际生活中的

real time 实时(在数据发生的当时处理该项数据,并在所需的响应时间以内获得必要的结果。)

ream [riːm] n. ;v. 铰孔,铰削

reamer ['ri:mə] n. 铰刀
reasoning ['ri:zəniŋ] n. 推理,推论;a. 理性的,推理的
recapture [ri:'kæptʃə] n. ;v. 取回,夺回,恢复,收复
receptive [ri'septiv] a. 接受的,有接受力的,易于接受的,容纳的
reciprocate [ri'siprəkeit] v. (使)往复运动
reciprocating [ri'siprəkeitiŋ] n. ;a. 往复(的),来回(的)
reclaim [ri'kleim] v. 回收,再生,重复使用
reclamation [reklə'meiʃən] n. 废料回收,再生,修整
reconfigurable 可重新配置,可重构
rectangle ['rektæŋgəl] n. 长方形,矩形
rectangular [rek'tæŋgjulə] a. 矩形的,成直角的
redefine [ri:di'fain] v. 重新定义
redesign [ˌri:di'zain] v. 重新设计;n. 新设计
redouble [ri'dʌbl] n. 再加倍,加强
redundant [ri'dʌndənt] a. 冗余的,过量的,重复的 n. 多余部分,备份
redundant system 冗余系统
reenter ['ri:'entə] v. 重新进入,重返大气层,重新加入
refine [ri'fain] v. 改善,改进
reflex ['ri:fleks] n. 反射,反应能力
regime [rei'ʒi:m] n. 方法,领域,范围
register ['redʒistə] n. 记录;v. 显示出,测量
regulating wheel (磨床上的)导轮
regulator ['regjuleitə] n. 调整器,调节器
reinforce [ri:in'fɔ:s] v. 加强,增强,强化
reinforced concrete 钢筋混凝土
reinvent [ri:in'vent] v. 从头开始重做某事(特别是不需要的或者无效率的努力)
reject ['ridʒekt] n. 等外品,下脚料,次品,拒绝,不合格品
relevant ['relivənt] a. 有关的,相应的
reliability [rilaiə'biliti] n. 可靠性,可靠度,安全性

remainder [ri'meində] n. 剩余部分

removal [ri'muːvəl] n. 除去,切削,切除

render ['rendə] v. 致使

rendering ['rendəriŋ] n. 透视图,示意图

renewal [ri'njuːəl] n. 更新,重新开始

renown [ri'naun] n. 名望,声望;v. 使有名望,使有声誉

renowned a. 有名望的,著名的

repeatability 可重复性,再现性

rephrase [riːfriz] v. 改用别的措词表达

replacement 替换,替换物

rerun [riː'rʌn] v. 重新运行

reside [rizaid] v. 驻留,居住,归于

resident in 归属于...的,存在于...中

resolution [rezə'ljuːʃən] n. 分辨率,分辨能力

resolve [ri'zɔlv] v. 分析,分辨,解决

resolver [riː'zɔlvə] n. 分解器(一种能将旋转的和线性的机械位移转换成模拟电信号的变换器)

respondent [ri'spɔndənt] a. 回答的;n. 回答者

responsiveness 反应性,响应性

rest button 支承钉

restrain [ris'trein] v. 抑制,约束,限制

restrictor valve 节流阀

resultant [ri'zʌltənt] a. 合成的,作为结果发生的;n. 生成物

retail ['reiːteil] n.;v. 零售

retool ['riːtuːl] v. 给(工厂、企业)以新装备,对机械进行改装(革新)以生产新产品

retrieve [ri'triːv] n.;v. 检索

return on investment 投资回报率

reversed [ri'vəːst] a. 颠倒的,反向的,相反的

revival [ri'vaivəl] n. 复兴,恢复,再生

revolutions per minute 转数/分,每分钟转数
revolving table 旋转工作台
rework ['ri:'wə:k] n. 重新加工,返工,修改
rib [rib] n. 肋,筋;v. 加肋,加筋
ridge [ridʒ] n. 隆起物
rigger ['ri:gə] n. 装配工
right hand thread 右旋螺纹
rigid coupling 刚性联轴器,刚性联接
rigidity [ri'dʒiditi] n. 刚性,刚度,坚固性
rigorous ['rigərəs] a. 严格的,严密的
robotics [rəu'bɔtiks] n. 机器人学,机器人技术
robust [rəu'bʌst] a. 健壮的,健全的,耐用的,坚固的
ROI (return on investment) 投资回报率
roster [rəustə] n. 名册,人名册
rotational inertia 转动惯量
rotor [rəutə] n. 转子,转动体
round pin 圆销
round-off 四舍五入,抹去零头
round-the-clock 不分昼夜的,连续不停的,连续二十四小时的
royalty ['rɔiəlti] n. 版权费,使用费
rule of thumb 单凭经验来做的方法,经验法则
rule-based 基于规则的
runout ['rʌn'aut] n. 偏斜,径向跳动

S

salability [seilə'biliti] n. 出售,畅销,销路,销路好
salespeople ['seilz,pi:pl] n. 售货员,销售人员
salient feature 显著的特征
sawing ['sɔ:iŋ] n. 锯,锯开,锯切
scale 比例尺,刻度尺

scale up (把)... 按比例增加,递加,增大比例
scan [skæn] v. 检查,扫描,浏览
schedule ['skedju:l] n. 时间表,进度表,预订计划
scheduling 编制作业进度计划
schematic [ski'mætik] a. 示意性的,图表的; n. 示意图,原理图,简图
schematic representation 图示,略图,简图,示意画法
schematically [ski'mætikli] ad. 用示意图,示意地
scheme [ski:m] n. 安排,配置,方案
scrap [skræp] n. 铁屑,切屑,边角料,废品,回炉料
scrapped 废弃的
screening 筛选
screw fastener 螺丝紧固件
sealing ['si:liŋ] n. 密封,封接,封口
seam [si:m] n. 接缝,焊缝,接口
seasoned ['si:zənd] a. 经验丰富的,老练的
secondary ['sekəndəri] a. 次要的,从属的,第二位的,间接的
secondary benefit 间接利益,次级收益
section 截面
segment ['segmənt] n. 线段,程序段
seize [si:z] v. (机器等)卡住,咬住,粘结
self-directed 自我指导的,自主的
self-exciting 自激的,自励的
self-locking [self'lɔkiŋ] a. 自锁的
seminar ['seminɑ:] n. 讲座,讨论会
semipermanent [ˌsemi'pə:mənənt] a. 半永久性的,暂时的
semiskilled worker 半技术工人,半熟练工人
senior executives 高级职员,高级主管人员
senior management 高级管理层
sensible ['sensəbl] a. 切合实际的,合理的,有判断力的
sensing 检测,感觉

sensor ['sensə] n. 传感器,传感元件

sentiment ['sentimənt] n. 感情,情绪,意见,感想

sequentially [si'kwenʃəli] ad. 顺序地

serial number 序号,编号

server 服务器

service of equipment 设备的技术维护

service sector 服务性部门

serviceability n. 使用(服务)能力,操作性能,维护保养方便性

servo ['sə:vəu] n. 伺服机构,伺服电动机,伺服传动装置

servovalve 伺服阀

set down 放下,使着陆

set up 装夹,安装

setting ['setiŋ] n. 环境,安装,设置值,设定值,安置

setup time 准备时间,辅助生产时间,建立时间,装工具时间

shaft [ʃɑ:ft] n. 轴,辊

shear [ʃiə] v. 剪切,切断

shed light on sth. 使某事清楚明白地显示出来,阐明

shelf life 保质期

shock [ʃɔk] n. 冲击,碰撞

shock load 冲击载荷,突加载荷

shop floor 车间,工厂里的生产区,生产区中的工人

shortcut ['ʃɔ:tkʌt] n. 近路,捷径,简化

shorthand ['ʃɔ:thænd] n. 简略的表示方法

shorthand notation 简化符号

short-duration 短期的

short-run 短期的,少量生产,短期生产

shutdown ['ʃʌtdaun] n. 关闭,断路,停止

side head 侧刀架

silicon carbide 碳化硅

silicon nitride 氮化硅

simplistic [sim'plistik] *a.* 过分简单的
simulation [simju'leiʃən] *n.* 模拟,仿真
sine wave 正弦波形
single-coil helical spring 单圈螺旋形弹簧
single-threaded screw 单头螺纹螺钉
sizeable ['saizəbl] *a.* 相当大的,大的
skeptical ['skeptikəl] *a.* 怀疑的,怀疑论的
sketch [sketʃ] *n.* 草图,略图;*v.* 绘草图
skilled worker 技术工人
slab [slæb] *n.* 板,板钢,板坯
slide 滑板(顶部与有关零、部件连接,底面具有导轨,可在相配合的零、部件上移动的部件)
slide projector 投影仪,幻灯片放映机
slider ['slaidə] *n.* 滑块(机构中与机架用移动副相连又与其他运动构件用转动副相连的构件)
sliding ['slaidiŋ] *n.*;*a.* 滑动,可相互移动
slippage ['slipidʒ] *n.* 滑动量,滑移,下降,动力传递损耗,转差率
slippery ['slipəri] *a.* 滑的,打滑的
slip-stick 滑动面粘附现象
small-lot production 小批量生产
Society of Manufacturing Engineers (SME) 美国制造工程师学会
software package 软件包,程序包
solder ['sɔːldə] *n.* 低温焊料,结合物;*v.* 低温焊接,软(锡,银,钎)焊
solid model 实体模型
solid modeling 实体造型
sound *a.* 完整的,坚固的,稳妥的,合理的,正确的,有根据的
space-age 太空时代
spatially ['speiʃəli] *ad.* 空间地,存在于空间地
SPC (statistical process control) 统计过程控制
specialist ['speʃəlist] *n.* 专家

specialization [ˌspeʃəlaiˈzeiʃən] n. 特殊化,专门化,专业
specialty [ˈspeʃəlti] n. 专长,专业(化),特制品,特殊产品
specification [spesifiˈkeiʃən] n. 规格,规范,技术要求,说明书,详细说明
spectrum [ˈspektrəm] n. 光谱,领域,范围,系列
speech synthesis and recognition 语音合成与识别
spillage [ˈspilidʒ] n. 溢出[溅出,倒出](的物质),泄漏
spindle [ˈspindl] n. 轴,主轴
spindle carrier 主轴鼓(内装若干根主轴,并能转位、定位的鼓轮部件)
spline [splain] n. 仿样,样条,花键,(pl.) 仿样函数,样条函数
spoilage [ˈspɔilidʒ] n. 损坏,变坏,废品,浪费的东西
spray painting 喷漆
spreadsheet 电子数据表,电子表格
square turret 车床四方转刀架
squirrel cage motor 鼠笼式电动机
stable state 稳定状态
staging [ˈsteidʒiŋ] n. 脚手架,工作平台,构架
stall [stɔːl] v. 失速,停车,停止转动,发生故障
stamping die 冲模
stamping press 冲床,冲压机
standard deviation 标准差
standardize [ˈstændədaiz] v. 标准化,规格化,统一标准
stark [stɑːk] a. 僵硬的,苛刻的
start from scratch 从头开始
start out 开始,着手进行
starter [ˈstɑːtə] n. 起动器,起动装置
startup = start-up 启动,开始工件,运转
state-of-the-art 技术发展水平,现代化的,科学发展状态
static [ˈstætik] a. 静力的;n. 静止状态
static stress 静应力

statistical [stə'tistikəl] a. 统计的,统计学的
statistical process control(SPC) 统计过程控制
statistics [stə'tistiks] n. 统计,统计学,统计数字
stator ['steitə] n. 定子
status quo 现状
steady state 稳定状态
steering wheel 方向盘
stellite ['stelait] n. 钨铬钴(硬质)合金
stepping ['stepiŋ] n. 步进,分级,分段
stepping motor 步进电机
sterile ['sterail] a. 枯燥无味的,缺乏独创性的
sticking ['stikiŋ] a. 粘的,有粘性的
stick-slip 蠕动,爬行
stiffness ['stifnis] n. 刚性,刚度
stochastic [stə'kæstik] a. 随机的,不确定的,偶然的
stop 挡块
storage tank 储罐,贮存罐
straight cut 直线切削
streamlined ['striːmlaind] a. 最新型的,改进型的
stress 强调,应力
stress relieving 应力消除热处理
strict liability 严格赔偿责任
stroke [strəuk] n. 行程
structural member 结构件,构件
stud [stʌd] n. 双头螺柱(两头均有螺纹的圆柱形紧固件)
subassembly [sʌbə'sembli] n. 部件,组件
subcontractor ['sʌbkən'træktə] n. 第二次转包,小承包商,分包工
subdivide ['sʌbdi'vaid] v. 细分,再划分,重分
subset ['sʌbset] n. 子集合,子系统,子设备
substitute ['sʌbstitjuːt] v. 代替;n. 代用品,代替者

subsume ['səb'sju:m] v. 包含,包括,把...归入(某一类)
subtlety ['sʌtliti] n. 微妙,微细,敏锐
sulphide ['sʌlfaid] n. 硫化物;v. 变成硫化物
summation sign 求和符号
superalloy [sju:pə'æloi] n. 超耐热合金,高温合金
supplier [sə'plaiə] n. 供应商,厂商
surface grinding 平面磨削
surface model 曲面模型
surface roughness 表面粗糙度
survey course 概论课
sustained [səs'teind] a. 持续的,不间断的
switchgear ['switʃgiə] n. 开关装置,配电装置
switching device 开关装置
synchronize ['siŋkrənaiz] v. 同步
synchronous ['siŋkrənəs] a. 同步的,同时发生的,同时出现的
syndrome ['sindrəum] n. 综合症,并发症
synergistic [sinə'dʒistik] a. 协同的,合作的,互相作用的
synthesis ['sinθisis] n. 合成,综合,拼合
synthetic [sin'θetik] a. 合成的,人造的;n. 合成物
synthetic lubricant 合成润滑剂
system integration 系统集成

T

tackle ['tækl] v. (着手)处理,从事,对付,解决
tailstock ['teilstɔk] n. 尾架,尾座
tangential [tæn'dʒənʃəl] a. 切线的,切向的
tangentially [tæn'dʒənʃəli] ad. 成切线地
tape punch 纸带穿孔
taper ['teipə] n. 锥度,圆锥
tapered ['teipəd] a. 锥形的

tapping [ˈtæpiŋ] n. 攻丝,攻螺纹
teach box 示教盒
teaching n. 教学,讲授,(复数)学说,主义,教导,(宗教)教义
team player 能共同努力和相互合作的人
teamwork 协作,协同工作
technology transfer 技术转让
teethe troubles 事情开始时的暂时困难,初期困难
teleoperator 遥控机器人,远程操纵器
temperature extremes 温度极限
template [ˈtempleit] n. 样板,模板,模型;v. 放样
tender [ˈtendə] n. 招标,投标
termination [tə:miˈneiʃən] n. 终止,结束
terminology [tə:miˈnɔlədʒi] n. 术语,专门名词
the weakest link 最薄弱的环节
thermal [ˈθə:məl] a. 热量的,热力的,热的
thermal conductivity 导热性,导热系数
thermal diffusivity 热扩散系数,温度扩散率
thermodynamics [ˈθə:məudaiˈnæmiks] n. 热力学
third-party 第三方的
threaded fastener 螺纹紧固件
throughput [θru:ˈput] n. 生产量,生产能力
through-hole 通孔
time-sharing 分时(当多用户通过终端设备同时使用一台计算机时,系统把时间分成许多极短的时间片,分配给每个联机作业,由时钟控制中断,使各作业交错使用计算机)
timing [ˈtaimiŋ] n. 定时,记时,时间控制
TIR (total indicator reading) 总读数,指针读数
tire [tiə] n. 层;v. 分层布置
titanium carbide 碳化钛
title block 标题栏

to true the wheel 对砂轮进行整形(产生确定的几何形状)
tolerance ['tɔlərəns] n. 公差;v. 给(机器零件等)规定公差
tolerancing charting 公差的图表计算法
tool [tu:l] n. 工具;v. 给...装备上工具,提供加工机械
tool and cutter grinder 工具磨床
tool angles 刀具角度
tool changer 刀具更换装置,换刀装置
tool magazine 刀具库
toolholder ['tu:lhəuldə] n. 刀夹(用来安装和紧固切刀的工具)
tooling 工艺装备,工装(产品制造中所用的各种工具的总称。包括刀具、夹具、量具、辅具等)
tooling library 工艺装备库
toolmaker 工具制造者
toolroom [tu:lrum] n. 工具室,工具车间
top management 最高管理部门,高层管理人员
torque [tɔ:k] n. 转矩,扭矩;v. 扭转
total quality management 全面质量管理
totality [təu'tæliti] n. 全体,总数,总额,完全
touch screen 触摸式屏幕
touch trigger 接触触发器
tracer ['treisə] n. 追踪装置,随动装置,仿形板
tracer lathe 靠模车床
track [træk] n. 轨道,磁道,导向装置;v. 跟踪,沿轨道行驶
trade deficit 贸易逆差
trade fair 商品交易会
trade journal 行业杂志
trademark 商标,标志,品种
tradeoff = trade-off 折衷(方案,方法),权衡
trade-in 折价,折价物;a. 折价的
training program 培训计划

trajectory ['trædʒiktəri] n. 轨迹,路线,路径

transcribe [træns'kraib] v. 抄写,记录

transfer line 自动生产线

transformer [træns'fɔ:mə] n. 变压器

transient ['trænziənt] a. 短暂的,瞬变的

transition [træn'siʒən] n. 转变,变换,过渡

transmission [trænz'miʃən] n. 传递,传动装置,变速箱

transmitter 变送器(输出为标准信号的传感器)

traverse ['trævə:s] v. 横向移动

tribological [ˌtraibəu'lɔdʒikəl] a. 摩擦学的

trim [trim] v. ;n. 使整齐,修整,调整,去毛刺

trouble-free 无故障的,可靠的

tungsten ['tʌŋstən] n. 钨

tuning ['tju:niʒ] n. 调整,调谐

turbine ['tə:bin] n. 涡轮机,透平(机)

turbulence ['tə:bjuləns] n. 湍流,旋涡

turnaround 转变,转向,突然好转

turnaround time 解题周期,周转时间

turning ['tə:niŋ] n. 旋转,车削

turning center 车削中心

turning machine 车床

turnkey system 整套系统,交钥匙系统

turnover ['tə:nəuvə] n. 人员调整,人员更新

turret ['tʌrit] n. 转塔,六角刀架,六角头

turret lathe 转塔车床

T-square 丁字尺

U

ultimate tensile strength 极限抗拉(拉伸)强度

UNC (unified coarse thread) 统一标准粗牙螺纹

uncommitted [ˌʌnkə'mitid] a. 自由的,不受约束的,不负义务的
unconventional [ˌʌnkən'venʃənl] a. 非传统的,不是常规的
undefined [ˌʌndi'faind] a. 未规定的,不明确的,模糊的
under consideration 在考虑中,在研究中
underdesign [ˌʌndədi'zaiŋ] n. 欠安全的设计
underestimate [ˌʌndər'estimeit] v. 低估,看轻
undergo [ˌʌndə'gəu] v. 经历,经受(变化)
underlying [ˌʌndə'laiŋ] a. 下面的,潜在的,根本的
underutilize [ˌʌndə'juːtilaiz] v. 未充分利用
underway [ˌʌndə'wei] a. 起步的,进行中的
UNEF (unified extra-fine thread) 统一标准超细牙螺纹
unemployment rate 失业率
UNF (unified fine thread) 统一标准细牙螺纹
unified standard 统一标准
uniformity [juːni'fɔːmiti] n. 同样,一致,均匀,一致性,均匀性
unintended [ˌʌnin'tendid] a. 不是故意的,无意识的
unlubricated [ʌn'luːbrikeitid] a. 无润滑的
unqualified [ˌʌn'kwɔlifaid] a. 不合格的,无条件的,绝对的
unquenchable [ʌn'kwentʃəbl] a. 不能熄灭的,不能遏制的
unscheduled [ʌn'ʃedjuːld] a. 事先未安排的,计划外的
unskilled worker 非技术工人,做粗活的工人
unsophisticated [ˌʌnsə'fistikeitid] a. 不复杂的,简单的
unstable [ʌn'steibl] a. 不牢固的,不稳定的
upper deviation 上偏差
up-front 在前面的,预先支付的,先期的
up-to-date [ˌʌptə'deit] a. 现代化的,最新的,尖端的
user-friendliness 用户友好的,容易使用的
utilise ['juːtilaiz] v. 利用,使用
utmost ['ʌtməust] n. 极限,最大可能的; a. 极度的

V

validation [væli'deiʃən] n. 使生效,合法化,批准
valve [vælv] n. 阀
vanishing ['væniʃiŋ] n. 消失
variability [vɛəriə'biliti] n. 易变性,变化性,变异度
variable ['vɛəriəbl] n. 变量;n. 变量的,可变的
variable cost 可变成本
variable expense 可变费用
variant ['vɛəriənt] a. 不同的;n. 派生,衍生
variant approach 派生法
vector ['vektə] n. 矢量,向量
vehicle n. 车辆,媒介物,运载工具,载体
vehicular [vi'hikjulə] a. 车辆的
verification [vərifi'keiʃən] n. 验证,确认,校验,检验
versatility [və:sə'tiliti] n. 通用性,多功能性,多方面适用性
versus ['və:səs] prep. 对...,与...相对
vertical mill 立式铣床
vertical turret lathe 立式转塔车床
vibration [vai'breiʃən] n. 振动
vibration-isolator 振动隔离器
vice versa ['vaisi'və:sə] ad. 反过来也是一样,反之亦然
view 视图
virtual company 虚拟公司
viscosity [vis'kɔsiti] n. 粘性,粘滞度
viscous ['viskəs] a. 粘的,粘性的,粘稠的
visualization [vizjuəlai'zeiʃən] n. 使看得见的
vitrified-bond 陶瓷结合剂
volume production 批量生产,成批生产

W

warranty ['wɔrənti] *n.* 保证(书),担保(书)

washer ['wɔʃə] *n.* 垫圈

wavy ['weivi] *a.* 波浪的,起伏的,有波浪的

ways and bed 导轨和床身

wear [wɛə] *v.* ;*n.* 磨损,损耗,磨蚀

wear-out 磨损,消耗,耗尽,用坏,用完

Web site 万维网站,万维站点

wedge [wedʒ] *n.* 楔,楔形物

weighing [weiiŋ] *n.* 权衡,权重

weld [weld] *n.* ;*v.* 焊接,熔接

what-if 假设分析,作假定推测

wheelhead 砂轮头

wheel-and-axle 轮轴

winding ['waindiŋ] *n.* 绕组,线圈

wire rope 钢缆,钢索,钢丝绳

wireframe model 线框模型

work 工件

work envelope 工作包络,加工包迹(表示机器人可达到的最大工作范围和作用距离的一些点的集合)

work head 转盘,工作台

workbench 工作台

working drawing 工作图(在产品生产过程中使用的图样,包括零件图和装配图)

workpiece ['wə:kpi:s] *n.* 工件

workshop ['wə:kʃɔp] *n.* 车间,专题研讨组,(专题)研讨会

workspeed = workpiece speed 工件速度

worktable 工作台

work-in-process 在制品(即在一个企业的生产过程中,正在进行加工、装配

或待进一步加工、装配或待检查验收的制品)
world-class 世界级的,世界水平的
worst-case approach 极值法
worthwhile ['wə:ð'hwail] a. 值得做的
wrench [rentʃ] n. ;v. 扳手,拧紧

Y

year in and year out 年年不断,始终不断地
yield strength 屈服强度

主要参考文献

[1] 施平. 机械工程专业英语教程[M]. 3版. 北京:电子工业出版社,2012
[2] 施平. 机械工程专业英语[M]. 15版. 哈尔滨:哈尔滨工业大学出版社,2014
[3] BENEDICT G F. Nontraditional Manufacturing Processes [M]. New York: Marcel Dekker,1987.
[4] BERTOLINE G R, WIEBE C L, MILLER C L, and NASMAN L O. Fundamentals of Graphics Communication [M]. Chicago: Irwin,1996.
[5] BLAKE A. Practical Stress Analysis in Engineering Design [M]. New York: Marcel Dekker Inc. ,1982.
[6] BRADLEY D A, DAWSON D, BURD N C, and LOADER A J. Mechatronics [M]. London: Chapman and Hall,1991.
[7] ECKHARTDT H D, Kinematic Design of Mechanics and Mechanisms [M]. New York: McGraw-Hill,1998.
[8] El WAKIL S D. Processes and Design for Manufacturing [M]. Englewood Cliffs: Prentice-Hall,1998.
[9] ESPOSITO A, THROWER J R. Machine Design [M]. Albany, NY: Delmar Publishers Inc. ,1991.
[10] GOETSCH D L. Advanced Manufacturing Technology [M]. Albany, NY: Delmar Publishers Inc. ,1990.
[11] GOETSCH D L. Modern Manufacturing Process [M]. Albany, NY: Delmar Publishers Inc. ,1991.
[12] HIGDON A, OHLSEN D H, STILES W B, WEESE J A, and RILEY W F. Mechanics of Materials [M]. New York: John Wiley & Sons,1985.
[13] KALPAKJIAN S, SCHMID S R. Manzrfacturing Engineering and Technology [M] Upper Saddle River, N. J. : Prentice Hall,2001.
[14] KOMACEK S A, LAWSON A E, HORTON A C. Manufacturing Technology [M]. New York: McGraw-Hill,1997
[15] KUSIAK A. Intelligent Manufacturing Systems [M]. Englewood Cliffs, NJ: Prentice Hall,1990.
[16] LINDBECK J R, WILLIAMS M W, WYGANT R W. Manufacturing Technology [M]. Englewood Cliffs, NJ: Prentice Hall,1990.

[17] MADSEN D A, SHUMAKER T M, TURPIN J L, and STARK C. Engineering Drawing and Design [M]. Albany, NY: Delmar Publishing, 1991.
[18] MATTSON M. CNC Programming Principles and Applications [M]. New York: Delmar, 2002.
[19] NIEBEL B W, DRAPER A B, WYSK R A. Modern Manufacturing Process Engineering [M]. New York: McGraw Hill, 1989.
[20] RAO P N, Manufacturing Technology Melal Cutting and Machine Tools [M]. New York: McGraw-Hill, 2000.
[21] REGH J A, KRAEBBER H W. Computer-Integrated Manufacluring [M]. 2nd ed. Upper Saddle River, NJ: Prentice Hall, 2001.
[22] SHIGLEY J E, MISCHKE C R. Mechanical Engineering Design [M]. New York: McGraw-Hill. , 1989.
[23] TRENT E M. Metal Cutting [M]. Oxford, UK: Butterworth-Heinemann Ltd. , 1991.